STO

June 28, 1978

Weather Modification in the Public Interest

Weather

Modification

in the

Public Interest

Robert G. Fleagle
James A. Crutchfield
Ralph W. Johnson
Mohamed F. Abdo

Published by the
AMERICAN METEOROLOGICAL SOCIETY
and the
UNIVERSITY OF WASHINGTON PRESS

Library of Congress Cataloging in Publication Data
Main entry under title:

Weather modification in the public interest.

    1. Weather control–United States.   I. Fleagle, Robert Guthrie, 1918–
II. American Meteorological Society.

QC928.7.W4   301.24'3   74–590
ISBN 0-295-95321-7

Foreword

As stated in the Preface, this report was prepared by a group of scientists at the University of Washington, in consultation with a number of persons from other organizations. It is intended to clarify a variety of policy issues relating to weather modification that must be faced here and abroad, and to provide additional background for those who must formulate these policies.

The American Meteorological Society is taking the initiative in distributing this report to a large number of key people in the U.S. Government, both Federal and State, so that they may act on the basis of an up-to-date summary of the state of weather modification research and the social and economic implications of future activities in this field.

The American Meteorological Society feels that this important document deserves a wide distribution in the public interest.

William W. Kellogg
President, AMS

October 1973

Preface

Serious problems of public policy relating to weather modification are coming into national focus. Policies developed in the next few years will likely have widespread impacts on the national welfare and, perhaps, on international relations. Clearly, the public as well as the scientific community and the executive and legislative agencies of government must be well informed on these issues. The authors hope that this monograph will contribute to the nation's quest for policies and programs in weather modification which are both technically effective and socially responsible.

In the development of weather modification technology our status is comparable to that of the early caveman who, watching a round boulder roll down a mountain slope, was inspired toward development of the wheel. He perhaps may have foreseen oxcarts, autos, highways, urban sprawl, air pollution, and the energy crisis, but it seems unlikely. We suggest that weather modification has reached a stage corresponding to development of the oxcart: It is not yet certain that there is a material hard enough to be used as the axle of the oxcart, and the question has not been seriously considered as to whether tribal unity will be jeopardized if wives and children are no longer needed as beasts of burden. Weather modification is faced with analogous problems which range from scientific and technical to economic, legal, and social.

The study reported here has been one of five parallel studies undertaken at the University of Washington over the past two years, by an interdisciplinary group of faculty and students, as part of an overall program directed toward understanding how society may improve the management of its technology. The five case studies concern the Highway Trust Fund, Hill–Burton hospital legislation, oil in the marine environment, nuclear power, and weather modification. Of these technologies, the last named is by far the newest and least institutionalized, and its future development is least certain. Consequently, opportunities may be greatest to apply to weather modification the lessons learned from histories of other technologies.

This report has benefited immeasurably from discussions with our colleagues in the Social Management of Technology project and with consultants who reviewed early drafts of the report. The counsel and criticism of the following persons have been especially appreciated: Peter V. Hobbs, Professor of Atmospheric Sciences, University of Washington; Milton Katz, Professor of Law, Harvard University; Thomas F. Malone, Dean, Graduate School, University of Connecticut; George E. Reedy, Dean, School of Journalism, Marquette University; Herbert R. Roback, Staff Director, House Government Operations Committee; Edward Wenk, Jr., Professor of Civil Engineering and Public Affairs, University of Washington; and Dael L. Wolfle, Professor of Public Affairs, University of Washington. The perspectives, conclusions, and recommendations are, of course, the responsibility of the authors alone.

The Social Management of Technology project, of which this study is an outgrowth, has been supported by the National Science Foundation under Grant No. GQ-2. The principal investigator of the project is Edward Wenk, Jr., Professor of Civil Engineering and Public Affairs. Co-investigators during most of the study period have been: James J. Best, Assistant Professor, Political Science; James A. Crutchfield, Professor, Economics; Robert W. Day, Professor and Chairman, Health Services; Robert G. Fleagle, Professor and Chairman, Atmospheric Sciences; Ronald Geballe, Professor and Chairman, Physics; Ralph W. Johnson, Professor, Law; John C. Narver, Professor, Business Administration; Dael L. Wolfle, Professor, Public Affairs. Mohamed F. Abdo has served as Research Assistant in the weather modification study.

March 1973   R. G. F.

Contents

Weather Modification in the Public Interest

I. Introduction

"We are faced with an insurmountable opportunity."

Pogo

Purpose of Study

Modification of weather by influencing the processes of cloud development is an emergent technology. So far, deliberate efforts to modify weather have had only marginal impacts on society, but activities of far greater import and potential benefit may develop in the future. These include modification of the amount or the distribution of rain, snow, and hail; dispersal of fog; and possibly the reduction of destructive hurricane winds. If large-scale operational programs are carried out in any of these areas, they will probably be accompanied by secondary effects which in many instances cannot be anticipated in detail, or even measured after the fact. These secondary effects will be very difficult to internalize within the technological system. In this respect weather modification offers an analogy to other environmental problems likely to assume increasing importance in the future: e.g., effects of air pollution on climate and on health, effects of pesticides and other toxic materials, effects of ocean pollution, and effects of changes in patterns of land use. In each of these fields, rational policy decisions are urgently needed to insure that activities are directed toward socially useful goals.

The purpose of this study is to appraise the steps, some of them rational and deliberate, some more or less accidental, which have been taken thus far in developing the capability of modifying weather, and to identify critical issues which limit development or which influence the ability to direct weather modification in a socially responsible manner. This appraisal indicates that the national effort has been deficient, especially in the areas of public policy and administrative management. A second objective of the study is, therefore, to design in broad outline an institutional means for rational, systematic examination of weather modification programs and for the development of policy and its implementation. It is obviously not possible to foresee all the dimensions of issues which may be encountered in weather modification, hence the proposed structure must be flexible and adaptive to new information and new perceptions as they arise, both in the determination of policy and in the administration of programs.

The study is directed specifically toward deliberate efforts to modify weather. Inadvertent effects on weather and climate caused by pollution or by changes in surface characteristics may present even more serious long-range problems which will be referred to in relevant context. We consider the formulation of rational policy for deliberate modification to be an essential first step toward coping with these less defined problems of inadvertent modification.

Recent studies of the scientific and technical aspects of weather modification, and others directed at the social and legal aspects, have been conducted by groups of experts working under the aegis of the National Academy of Sciences or under grants or contracts by government agencies. For at least three specific operational projects, extensive technological assessments have been made during the past two years. The present study utilizes each of these prior efforts; however, its fundamental perspectives are substantially unique in contending (1) that sound policy in weather modification requires continuing and thoroughly integrated attention to issues of public policy arising out of the scientific, socioeconomic, legal, and administrative aspects of the subject; and (2) that existing institutional mechanisms may not be able to respond appropriately to future problems in weather modification.

Basic Attitudes toward Weather Modification

Man lives at the bottom of an ocean of air, his welfare dependent on its life-sustaining beneficence. However, the atmosphere's hostile moods are an ever-present threat, and man must constantly pit his intelligence against the buffeting of storm, the vicissitudes of rainfall and drought, and the rigors of paralyzing cold. No aspect of his environment affects him more directly or more constantly than the atmosphere and the weather it generates. Small wonder that from earliest times man has identified changes in weather with awesome and myterious forces. Legends and historical records show how he first sought to placate and later to control or modify these forces.

Today, man has achieved capabilities to modify weather based on newly won understanding of atmospheric processes and new technical capacities. Great uncertainties remain, however, and atmospheric scientists differ regarding the magnitude of possible effects. The layman, having heard widely varying claims, has little basis for discriminating judgment. One man automatically believes reports of success, as perhaps in Stone Age times he would have believed predictions of the medicine man. Another distrusts and resents all efforts to tamper with the weather, as perhaps in an earlier epoch he would have felt rain dances a presumptuous affront to the ruling deity. So even small efforts to modify weather in specific and limited ways evoke responses which are linked to a vast sounding board resonating with the overtones of man's basic attitudes toward natural events. Every man believes that he has a stake in efforts to modify weather; his responses are conditioned by a host of factors ranging from scientific evidence through economic self-interest to subconsious attitudes and prejudice.

Those who live in North America have special reason to be sensitive to weather modification. Our climate is highly varied; rapid and dramatic changes in weather are common over much of the country— more so than in Western Europe, for example. A familiar observation could well refer to St. Louis, or Boston, or Seattle: "If you don't like our weather, just wait a few minutes." Few weeks go by without a major news story of flood, tornado, destructive cold, drought, or hurricane somewhere in the United States. A single storm can cost billions of dollars in damage, while tornadoes cause more than 100 deaths in the average year and property losses of about $75 million. Destruction due to hurricanes, hail, floods, droughts, or forest fires caused by weather conditions is often of major proportions; however, no reliable and complete accounting of losses associated with all types of weather events has ever been compiled.

In this country attitudes toward weather modification have also been shaped by the history of exploitation of the land and forests and especially the development of irrigation projects in the West. Until quite recently government policies have encouraged the unrestrained exploitation of whatever nature provided: minerals, forests, land, water, or clean air. Since these policies have come to seem reasonable and natural to those who benefited from them, it is understandable that the weather modification laws of many Western states treat the water contained in clouds as analogous to minerals under the land or water flowing over it. Other persons, more sensitive to their responsibilities to future generations, have created a powerful environmental lobby to oppose the exploitation of nature and have fostered the development of a mystique of conservation which, by extension, contributes to popular distrust of weather modification. The laws of several Eastern states forbid or severely restrict all efforts to modify weather.

Basic attitudes toward weather modification thus vary from person to person and from region to region. It is clear that the public interest is heavily involved in the future of weather modification and that capabilities must be developed within policies which are socially responsible. Decisions must be based on consideration of a broad range of issues, technical and scientific, economic, legal, and social. The distinction between research and operations will be critically important in all decisions that shape the future of weather modification.

Sensitivity to Weather

Technology has enabled many people to nearly ignore changes in weather. It is easily overlooked, therefore, that the sensitivity of the social system to weather events actually is increasing from year to year. Economic losses due to hurricanes, floods, tornadoes, freezes, blizzards, drought, and smog steadily increase with increased population and property development. For example, from 1915 to 1934 hurricane damage to U.S. coastal property averaged about $40 million annually, adjusted to construction costs based on the years 1957-59, whereas in the period from 1965 to 1970 damage averaged about $500 million annually on the same base.[1] This represents an increase at the rate of about 6 percent per year.

As technological systems and the organization of society become more complex, vulnerability to weather-caused disasters increases drastically. Widespread interruptions in electrical power, which once could be tolerated easily, now cause catastrophic disruptions. This example can easily be multiplied many times in the fields of transportation, food distribution, communications, and other vital services.

As the number and complexity of interactions of society with its environment predictably continue to increase, the possibilities for effective weather modification are likely to appear more and more attractive, and we must be able to discriminate in each case whether the promise is real or illusory. The issues that confront us now will not vanish of their own accord; they are likely to escalate as the stakes get larger.

Physical Bases of Efforts to Modify Weather

The weather modification efforts discussed in this study are based largely on the Bergeron-Findeisen process of ice crystal growth. This process depends critically on the fact that at temperatures colder than the melting point of ice ($0°$ Celsius) clouds typically contain both large numbers of supercooled liquid droplets and a certain number of ice crystals. The vapor pressure over the droplets is higher than that over the ice crystals at the same temperatures. Consequently, water vapor is deposited on ice surfaces while it evaporates from liquid surfaces, and the net effect is to transfer water rapidly from droplets to crystals. When the ice crystals reach precipitable size, they fall and coagulate with supercooled droplets and other crystals, in some cases reaching the ground as snow, in other cases melting as they encounter warmer air near the earth. This is the process responsible for much of the rain and snow that falls in middle and high latitudes.

The creation of ice crystals within clouds is thought to depend essentially upon the presence of preexisting ice crystals or on the presence of minute solid crystalline particles called *ice nuclei*. Clouds containing few ice nuclei may reach $-20°C$ or lower temperatures without the droplets freezing. The addition of artificial ice nuclei to such a cloud may trigger the Bergeron–Findeisen process and produce more snow or rain than would have occurred naturally. The seeding agent may be microscopic particles of silver iodide (AgI), dry ice (solid CO_2), or other materials.

At temperatures above $0°C$ clouds produce precipitation through the transfer of water vapor to droplets and by coalescence of droplets. Small water-soluble particles, called *cloud condensation nuclei*, play crucial roles in the growth of droplets. Although nearly always the supply of nuclei is ample to initiate droplet formation at relative humidities above 100 percent, the growth of droplets can be influenced by the addition of soluble particles. Finely ground sodium chloride (common salt) is often used.

A variety of effects may be produced in a cloud depending on many factors including the concentration of natural ice nuclei and cloud condensation nuclei, concentration and sizes of droplets, temperature, and the rate at which water vapor is supplied to the cloud by updrafts. Effects will depend also on the number of artificial nuclei added per unit volume and their physical properties. Moisture-laden clouds can under favorable circumstances be made to release water earlier and to precipitate at a greater rate than normal, thereby increasing the rainfall over dry regions. Snow clouds that form on the windward side of mountains can be encouraged to precipitate, thus contributing to a deeper winter snowpack than would occur naturally.

Cloud seeding may also result in decreased precipitation. If very large numbers of ice nuclei are introduced into a supercooled cloud, the resultant great numbers of ice crystals will compete for the available water vapor and thus few will grow to precipitable size. This process is referred to as *overseeding* and the cloud as *glaciated*. In some cases seeding may produce smaller snowflakes or a different crystal geometry which changes the pattern of fall, thereby causing redistribution of snowfall. By such techniques, the natural heavy concentration of snow downwind of the Great Lakes might be spread farther inland over larger areas. Addition of large numbers of cloud condensation nuclei to clouds above $0^{\circ}C$ may also result in decreased precipitation.

The latent heat released when water droplets change to ice crystals expands the air in the cloud and, under certain circumstances, increases buoyancy. In this way seeding may result in vertical growth of clouds to greater heights than they would reach naturally. Expansion of air within the cloud also results in fall of atmospheric pressure in the column below the cloud. If seeding is carried out adroitly, the reduced pressure may affect wind speeds in a desired manner; conceivably, the violent winds of hurricanes and other severe storms might be reduced in this manner.

Supercooled fog over airports or highways can be dissipated by adding ice nuclei which initiate the growth and precipitation of ice crystals at the expense of the many smaller droplets. Hailstorm damage may be reduced by adding artificial ice nuclei, thereby stimulating the precipitation of many smaller hailstones instead of the fewer larger ones which occur in hail clouds.

The processes of cloud development and precipitation, both natural and modified, occur as parts of larger-scale atmospheric processes. It is possible to modify clouds to a significant degree only in those cases where the preexisting large-scale atmospheric processes are favorable. It is important, therefore, that research in weather modification be closely associated with research in storm generation, air–sea interaction, the general circulation, and other aspects of the total field of meteorology. Operational programs in weather modification necessarily should be designed and carried out on the basis of broad understanding of atmospheric processes responsible for weather. In fact, weather prediction and weather modification should be recognized as essentially coupled. Improved capability for prediction may open the way to modification, and modification should generally be attempted only if its effects can be predicted accurately and reliably.

We must recognize a range of capabilities to predict and to control phenomena, and the strategies adopted should vary accordingly. The following different situations should be anticipated.

1.   Direct, effective control; little uncertainty in prediction.
2.   Some degree of modification reasonably certain, but considerable uncertainty as to amounts or location of precipitation, etc.; prediction statistics good, but individual cases may be unsuccessful.
3.   Redistributive effects of modification so large, or uncertainties so large, that modification efforts are unwarranted; efforts should be concentrated on prediction.

These different categories obviously impose different requirements with respect to research, operations, and the decision-making process.

The physical processes associated with inadvertent modification are widespread and virtually uncontrolled, and understanding of important aspects remains at a primitive level. The impact on society is more extensive and, in the long run, may be more serious than the consequences of deliberate modification. In certain respects deliberate and inadvertent modification can be usefully considered together, as for example in urban or industrialized areas where pollutant gases and particles may produce effects in clouds similar to the results of deliberate seeding. Fundamental studies of cloud physics are equally essential in either situation, and what we learn in pursuit of one goal will be valuable to the other, both in the area of science and in the area of public policy.

In other respects inadvertent modification presents physical problems which extend well beyond those of deliberate modification. One such problem concerns the effects of pollutants on the radiation absorbed and reflected by the atmosphere; this is basic to an understanding of how the climate may change due to air pollution. Another problem concerns the acidity of rainfall and its distribution downwind from sources of air pollution. Still another stems from the possible effects of a fleet of supersonic transports on the ozone concentration of the stratosphere and its implications for the health of the general public. Clearly, serious public policy issues are associated with these problems.

Emergence of Issues

In 1946 Vincent Schaefer, a research scientist for the General Electric Corporation, demonstrated that solid CO_2 (dry ice) dropped into supercooled clouds can initiate the rapid transformation of a cloud of supercooled water droplets into ice crystals. In 1947 Bernard Vonnegut of the same laboratory demonstrated a similar effect using silver iodide particles. These experiments were carried out under the general direction and with the enthusiastic acclaim of Nobel Laureate Irving Langmuir, who concluded that man's control of large weather systems was imminent. Although many atmospheric scientists remained skeptical about the evidence, Langmuir was an effective spokesman, and the public was generally quite prepared in the wake of great wartime technology successes to accept even sweeping claims of success in modifying weather.

If the public statements of Langmuir and other enthusiasts were to be believed, rich rewards awaited those able to capitalize on the opportunities. Numerous cloud-seeding operations were undertaken, until by the early 1950's 10 percent of the land area of the United States was under commercial seeding operations. The annual investment by ranchers, orchardists, towns, public utilities, and resort owners reached $3 million to $5 million.[2] Many operations were based more on aggressive public relations than on professional knowledge and skill, and antagonisms developed between the scientific community, which recognized that the potential benefits of weather modification had not yet been demonstrated, and the commercial operators, many of whom found it easier to exhort than to investigate.

The U.S. Weather Bureau began research on weather modification in 1948 under the direction of Dr. Ross Gunn. In the eyes of Weather

Bureau scientists familiar with the natural variability of weather, the results of the first comprehensive field tests did not substantiate Langmuir's startling claims. Viewpoints became strongly polarized between the General Electric group and the Weather Bureau. The Senate Interior Committee, hoping for increased water supply to the arid Western states, supported increased research in weather modification and also provided a platform for some of the more extravagant enthusiasts, whereas a large part of the scientific community, critical of the evidence, urged further research at a modest level. Although the Weather Bureau played a responsible role in these events, it was forced into the position of critic, which underscored its reputation as an old, overconservative bureaucracy.

Out of the controversy was born the President's Advisory Committee on Weather Control which was charged in 1953 by Public Law 83-256 with evaluating the results of experiments and with "determining the extent to which the United States should experiment with, engage in, or regulate activities designed to control weather conditions." The orientation of the Committee was decidedly *pro* weather modification, and this view was transmitted to the Senate Interior Committee through Lewis Douglas, a respected public member of the Advisory Committee who had the confidence of Senator Clinton Anderson, Chairman of the Senate Committee. The Advisory Committee reported in 1957 that seeding was effective in producing an average increase in precipitation of 10 to 15 percent in mountainous areas of western United States in winter, and it recommended that the National Science Foundation (NSF) be assigned primary responsibility for support of research in cloud physics and weather modification.[2]

As recommended by the Committee, Congress in 1958 assigned lead agency responsibility to NSF. However, because NSF was more responsive to the scientists' viewpoint than to the operators or the Interior Committee, the NSF program was pitched at a conservative funding level and emphasized laboratory research in universities rather than field experiments in cloud seeding. The fact that research was controlled by scientists proved a source of irritation to commercial operators, the Interior Committee, and the Interior Department during succeeding years. Beginning in 1959, NSF compiled an annual report of weather modification activities. NSF also supported a RAND Corporation study which in 1962 placed weather modification in the context of the broader field of atmospheric research and urged a definitive test of the modification of orographic clouds.[3]

Commercial cloud seeding during the early 1950s proved to be disappointing to those paying the bills, and by the end of the decade operations were down to about 10 percent of the highest level reached ten years earlier. Some operational programs, notably those of the Pacific Gas and Electric Company in California, continued through this period. Accumulated experience and results of the federally supported research program have now brought about a second phase of growth; NSF reported that in 1965 commercial operations in the United States covered 98,000 square miles in 26 states, 3 percent of the land area of the country.[4]

Cloud seeding has also been used in efforts to diminish the maximum surface winds occurring outside the eye of hurricanes. Project Stormfury, a joint project of the Commerce and Defense Departments, has sought since 1960 to develop methods for hurricane wind control. Numerical models have been used in designing possible modification efforts. Although only four field trials have been made in the twelve years of the project, the results have been sufficiently encouraging that current plans call for increased effort toward modification of hurricanes and other severe storms in the next few years.

During the 1960s two controversies dominated discussions of weather modification. The first concerned the difficult problem of separating the effects of seeding from the natural variability of storms. The 1957 report of the Advisory Committee had been strongly condemned by several statisticians for lack of proper randomization of the cloud-seeding trials. During the early 1960s only a few randomized experiments were conducted, so that in reviewing the subject in 1966 the National Academy of Sciences Committee on Atmospheric Sciences (NAS/CAS) found it necessary to rely largely on analysis of nonrandomized operational data. The Committee report[5] expressed cautious optimism for enhancing orographic precipitation (produced by air flow over mountains). Soon after publication this report was strongly attacked by the statisticians. This diverted attention from the substantive issue, but the criticisms did result in much more careful attention to statistical design of subsequent field experiments and to careful statistical analysis of results. The NAS report marked a turning point in the attitude of the scientific community and of government toward weather modification. Attracting the serious attention of scientists, economists, statisticians, lawyers, and government administrators, the NAS report laid the basis for expansion of federally supported research programs. The results from randomized experiments summarized by a more recent report of the Committee on Atmospheric Sciences[6] indicate that seeding may increase or decrease precipitation from orographic clouds and that the nature of the results depends upon measurable properties of the atmosphere. Also in early 1966, reports of the NSF Special Commission on Weather Modification[7] and of the NSF-supported Symposium on the Economic and Social Aspects of Weather Modification[8] provided the first detailed discussions of possible broad impacts of weather modification programs and emphasized the need for further consideration of related problems.

Principals in the second controversy have been the Departments of Commerce and Interior, with the NSF an interested observer who occasionally has been drawn into the fight. The substantive issue has been whether or not weather modification capabilities justify federal support for operational programs to augment precipitation. Since 1964 Interior has persistently expressed a strong affirmative and sought the dominant role in weather modification; Commerce has felt that the focus of effort should remain on research and has sought to develop its capability in both laboratory and field research. Two bills reflecting these points of view were introduced into the Senate in January 1966, respectively by Senator Anderson for the Interior and by Senator Magnuson for the Commerce Committees. Although the Magnuson bill passed the Senate in 1967 it did not become law. The provisions of these two bills and the history leading to their demise have been reviewed by R. W. Johnson.[9]

Within the Executive Branch the controversy was centered in the Interdepartmental Committee for Atmospheric Sciences (ICAS) which in May 1967 recommended that Commerce be assigned lead responsibility for developing a broad program of research but at the same time recommended that Interior's weather modification budget should increase to $35 million by FY 1970, a sum larger than that of any other agency. In 1968 a backward half-step was taken with the passage of Public Law 90-407 which removed the NSF mandate for coordination of weather modification, left coordination to the Executive Office of the President (EOP) and ICAS, and encouraged a variety of agency missions in addition to research. In June 1971, ICAS designated seven of the existing agency programs as national projects and encouraged accelerated effort to (1) augment the Colorado River Basin snowpack (Interior); (2) reduce hurricane winds (Commerce); (3) reduce fires caused by lightning (Agriculture); (4) increase rainfall from cumulus clouds (Commerce); (5) reduce hail damage (NSF); (6) spread snowfall inland from the Great Lakes (Commerce); (7) improve visibility in fog (Transportation).[10] In December 1971 passage of Public Law 92-205 required that reports of nonfederal weather modification efforts be filed with the Department of Commerce, and an administrative action is under consideration which may extend this requirement to include civilian federal agency programs. Certain Defense Department programs probably will be reported, but other large operational military programs reputedly have been conducted secretly.[11, 12] So difficulties stand in the way of complete reporting of weather modification activities.

Under the 1969 National Environmental Policy Act (Public Law 91-190) impact statements must be filed for weather modification programs. In response to this legal requirement and to the mounting public concern with environmental problems, studies have been made recently of certain aspects of the impacts of the following programs: augmentation of the Colorado snowpack, hurricane modification, precipitation modification, and airport fog dispersal. These studies in sum suggest the breadth of considerations that must be considered in future weather modification programs.

In the past several years extensive cloud seeding by aircraft has attempted to supplement natural rainfall in drought-stricken areas in the Philippines (1969) and in Texas, Oklahoma, and Florida (1971). These programs, carried out under U.S. federal auspices at the request of the local governments concerned, have caused minimal undesirable impacts. But it is easy to foresee the occasion in which a dry season that appears as a disastrous drought to one state may well be acclaimed as a boon to the tourist season of a neighboring state. The mechanism of government is not adequately equipped to cope with such issues.

The international dimensions of weather modification conceivably may present still greater problems. Every nation must be concerned about possible downwind effects of programs carried out by another nation. Most critical are the implications of possible large-scale and secret efforts to modify weather for aggressive military purposes. No convincing evidence is available that suggests any major effects, but the possibility cannot be ruled out that far-reaching deleterious effects might still be produced. The Department of Defense has been interested in military applications since the early 1950s, and rumors of military operations in southeast Asia have circulated since 1967. Recent articles in *Science* and the *New York Times* review what is known about these activities outside the limits of classified information and indicate that a major international issue may be developing.[11, 12]

Finally, the problems of inadvertent modification cited earlier present an additional set of public policy issues. These include effects of air pollution on cloud and precipitation processes and on the composition of rain, effects of the radiation absorbed and emitted by the atmosphere, and effects of aircraft on the ozone concentration of the stratosphere. Local effects of pollution on cloud processes are so closely related to the effects of cloud seeding that these specific effects probably should be handled together. Effects in the other areas are so widespread and so little understood that it appears most appropriate to handle them separately. They are therefore not addressed explicitly here, although this study is regarded as an initial step toward consideration of the issues arising from inadvertent modification. Public policy and administrative mechanisms developed for deliberate modification may well ultimately form a basis for overall prudent management of our atmospheric environment.

The issues we now face in weather modification have roots in the science and technology of the subject, but no less importantly in the politics of government agencies and Congressional committees and in public attitudes which grow out of a variety of historical, economic, and sociological factors. In summary, the most critical developments that have shaped the issues we now face in weather modification are listed in Table 1.

Two major issues of long standing have been essentially resolved within the past several years, at least as they apply to the scientific community. The approximate magnitude of the increases in precipitation to be expected as a result of seeding orographic clouds has been determined within broad but useful limits. We can now talk with some confidence about potential increases from 10 percent to 50 percent or more over that which would occur naturally, and about decreases of comparable magnitudes. Ten years ago, objective evidence was inadequate to establish even the order of magnitude of effects of cloud seeding.

The second issue that has been substantially resolved in recent years concerns the emphasis to be placed respectively on research or on operations. As uncertainties have been defined and as experience has accumulated, the need for research has been more widely acknowledged in all agencies of government as well as throughout the scientific community.

Other vital issues that will determine the future of weather modification remain to be resolved. Among them are the following:

Table 1

Critical Developments Affecting
Weather Modification Policy

Event	Date	Effects
Schaefer-Vonnegut experiments	1946-47	Demonstrated technical possibilities.
Langmuir advocacy	1947-55	Gained wide public interest, stimulated operational programs.
Field experiments (Project Cirrus, ACN)	1947-53	Demonstrated limitations.
Report of Pres. Advis. Comm.	1957	Clarified evidence and placed focus on need for research.
NSF Research Investment	1958-present	Developed research groups and provided research base.
Dept. Interior Initiative	1964-present	Advanced case for operational cloud seeding.
Project Stormfury	1960-present	Developed possibility of severe storm modification.
NAS/CAS Report (1350)	1966	Laid basis for expanded federal research programs.
Randomized Field experiments	1962-70	Established magnitudes of orographic seeding effects.
Nat. Environ. Policy Act	1969	Directed attention toward broad impacts.
Military Operations	1967-present	Raised issue of military use.

1. Answers to scientific questions requiring substantial resources for their solution; these include (a) determination of whether winds in hurricanes and other severe storms can be reduced in severity, and (b) more precise determination of conditions under which precipitation can be increased or decreased.

2. Decisions on a lead agency responsible for coordinating the federal programs in weather modification and for carrying out research requiring large resources of manpower and facilities.

3. Objective, reliable assessments of the economic and sociological impacts of weather modification programs.

4. Development of uniform and appropriate federal regulations and legal provisions for liability.

5. Development of a mechanism for investigating public policy issues on a continuing basis, and for providing sound policy initiatives.

6. Consideration of possible international use of weather modification as a weapon of war.

The issues of weather modification should not be seen as isolated or self-contained. Rather, they are part of the large problem area which can be called environmental management. For example, viewed in this way the pertinent question regarding hurricanes is: Can damage and loss of life best be minimized in a socially responsible manner by efforts toward improved prediction, by modification, or by a combination of both? Analogous questions may be asked for drought, fog, flood, hail, urban air pollution, and other important weather phenomena. Obviously, appropriate decisions should be expected to vary from case to case. Cold fog may be handled by direct, controlled use of modification techniques, whereas drought might best be coped with in a much more complicated way—for example, by a combination of modification of orographic precipitation within a limited area, recycling of used water, and shift in agricultural use of portions of the land area. The optimum combination is likely to change with time as prediction and modification improve and uncertainties are reduced.

Current Capabilities

Weather modification remains a controversial field. Scientific understanding has progressed in some areas more rapidly than in others, and it appears that the various forms of modification discussed earlier in this section are by no means equally effective.

Current capabilities have been appraised recently by the Panel on

Weather Modification of the National Academy of Sciences Committee on Atmospheric Sciences.[6] This report summarizes present capabilities for modifying precipitation in the following words: "On the basis of statistical analysis of well-designed field experiments—ice-nuclei seeding can sometimes lead to more precipitation, sometimes lead to less precipitation, and at other times the nuclei have no effect, depending on the meteorological conditions. Recent evidence has suggested that it is possible to specify those micro- and meso-physical properties of some cloud systems which determine their behavior following artificial nucleation." More specifically, with respect to orographic clouds, the report concludes: "In the longest randomized cloud seeding research project in the United States, involving cold orographic winter clouds, it has been demonstrated that precipitation can be increased by substantial amounts and on a determinate basis." With respect to convective clouds the report states: "The recent demonstration of both positive and negative treatment effects from seeding convective clouds emphasizes the complexity of the processes involved. They indicate that a more careful search must be made to determine the seedability criteria that apply to the convective clouds over various climatic regions. The economical exploitation of these weather modification techniques depends to a large measure on the development of such seedability criteria based upon the dynamics and physics of convective precipitation."

Concerning the seeding of large storm systems the report concludes: "Although some researchers claim that it is possible to increase and redistribute precipitation from certain nonorographic and nonconvective clouds, further experimental evidence is needed before any firm conclusions can be reached.... There is a pressing need for further analyses of the aerial extent of seeding effects under a variety of meteorological and topographical situations, and investigations into the physical mechanisms which are responsible for any such effects."

Concerning dissipation of fog, the report concludes: "The dissipation of supercooled fogs and low stratus clouds over limited areas is operationally practicable.... A fully reliable and practical procedure for dissipating warm fogs over airports still does not exist."

The conclusion with regard to lightning suppression is that "more experiments will be required." Possibilities of hail suppression are summarized: "There is a wide range of opinion on whether or not hail can be effectively suppressed or its damage mitigated.... The claims have sufficient physical plausibility and the reported results are sufficiently striking to conclude that for the realization of the possibility of ameliorating hail damage to agricultural crops and property, well designed and controlled hail modification experiments should be conducted."

On the subject of hurricane modification the report states: "It cannot yet be decided whether it is possible to reduce the intensity of winds in hurricanes by means of large doses of artificial ice nuclei. Nevertheless, the few observations available and the predictions from some numerical models are encouraging. Therefore, in view of the considerable potential benefits of success in moderating the destructive forces of hurricanes, we conclude that experimental and theoretical work on this problem should continue." With respect to modifying tornadoes and other severe storms: "It is concluded that a vigorous program of tornado research should be mounted with the long-range goal of developing an effective means of mitigating the destruction of these storms."

II. Major Programs

> "He had been eight years upon a project for extracting sun-beams out of cucumbers, which were to be put into vials hermetically sealed, and let out to warm the air in inclement summers."
>
> Jonathan Swift

Introduction

Federal weather modification programs have expanded markedly in the past decade, but private commercial operations have remained at a relatively low level. Federal activities shared among some ten agencies now dominate the field and are likely to continue to do so. The variety in objectives pursued is consistent with the missions of the individual agencies. There is little duplication of effort, due more to the independence of agency missions than to rational planning or central management. Efforts at effective coordination have been handicapped by interagency rivalry and administrative emphasis on agency survival at all cost. Passage of Public Laws 90-407 in 1968 and 92-205 in 1971 has left responsibility for coordination of weather modification divided between the Interdepartmental Committee for the Atmospheric Sciences (ICAS) and the Department of Commerce. ICAS is an effective mechanism for interagency discussion and for information exchange, but it has proved ineffective in establishing policy, initiating new research programs, or influencing programs developed by the separate agencies.

In 1971 ICAS proposed that seven agency programs be designated as National Projects, lead agencies were identified for each, and sub-stantial increases in funding for weather modification were recom-mended.[10] The Federal Council for Science and Technology approved these national projects on July 27, 1971. Each lead agency is respon-sible for coordination of its designated project, but the total program is linked only loosely through ICAS as before. The decision imple-menting the national projects represents an interagency agreement to do nothing at this time about designating a single lead agency.

The national projects constitute those elements of the total federal research program most likely to lead to operational programs. As they are described below, together with accounts of a few other related projects, the important component of research carried on by individuals and small groups in the U.S. and in other countries is not fully visible. These programs, conducted primarily at universities and at the National Center for Atmospheric Research (NCAR), have achieved most of the progress in understanding cloud processes and are vital in sustaining the research base and in training young scientists. Universities and NCAR also participate directly in some of the national projects.

Major Projects

To discuss the major projects efficiently, it is helpful to organize them under the general headings: Precipitation Augmentation and Redistri-bution; Reduction of Severe Storm Winds; Reduction of Hail, Light-ning, and Fog. It should be understood, however, that the projects themselves are not so organized.

Precipitation Augmentation and Redistribution

Colorado River Basin Project. Directed and supported by the Bureau of Reclamation of the Department of Interior, this is a three- to five-year pilot project to develop effective operational techniques for increasing winter snowfall in mountain areas. The study will assess downwind effects and economic, social, and ecological impacts. The target area of 2200 square miles is located in the San Juan Mountains of Colorado. The plan is to seed selectively only on days when cloud-top temperatures are between about -12°C and -23°C, the conditions most favorable for increasing snowfall according to results obtained in the Climax I and II experiments conducted by Colorado State University.[13] Seeding will not be conducted on days when the cloud-top temperatures are below about -23°C because previous experiments show that this would be likely to result in decreased snowfall. The project will involve six federal departments and agencies, state agencies, and three universities. Seeding operations and data collection will be carried out by private corporations.

Based on the results of earlier experiments, the Bureau of Reclamation estimates that seeding on all suitable days should result in a 30 percent increase in snowpack for an average winter season and should add as much as $100 million to the value of the stored water, thirty times the anticipated cost of the program.[10] However, both the benefits and costs remain controversial and uncertain, and other estimates of benefit-to-cost ratios are much more modest. In view of the uncertainties, introduction of estimated benefit-to-cost ratios at this time may invite misunderstanding. This is one of fifteen pilot and experimental research programs which constitute the Atmospheric Water Resources Program (Skywater) supported by the Bureau of Reclamation.

High Plains Project. The Office of Weather Modification of the Environmental Research Laboratories of NOAA developed plans in1972 for the High Plains Precipitation Enhancement Research Project.[14] Although not designated as a national project, this operation has been planned as a year-round program to be carried out at a high-plains site still to be selected. The project plan includes model studies, cloud and field experiments. Plans call for evaluation of results in the target area and in areas outside the project, and evaluation of economic, social, ecological, and legal impacts. Private industry and universities are to do much of the work under contract. In early 1973, the High Plains Project was substantially scaled down by direction of the Office of Management and Budget and responsibility shifted to the Bureau of Reclamation. It is not yet clear what the effects of this change will be.

Cumulus Modification Project. This project is directed by the Experimental Meteorology Laboratory, a division of the Environmental Research Laboratories of the National Oceanic and Atmospheric Administration (NOAA) and is located in Miami. The project seeks to develop means for increasing the buoyancy of cumulus clouds, thereby increasing their vertical growth and the rainfall which they produce. Mathematical models are developed and used to investigate the optimum plan for seeding, which is then carried out by aircraft and the results measured by radar and by aircraft. Field experiments have been successful in bringing about the merging of individual cumulus clouds and in increasing their growth and the rain produced. The results obtained so far do not provide adequate statistical data on which to design operational programs, but local political pressures led to an attempt to alleviate the south Florida drought in 1971.

Great Lakes Snow Distribution Project. This national project, conducted from 1968 to 1973 under the auspices of Environmental Research Laboratories of NOAA, was designed to shift some of the winter snowfall from the southern shore of Lake Erie farther inland to areas which normally receive less snow. North winds blowing across the Great Lakes in winter pick up moisture from the relatively warm lake surface; as the air is lifted over the downwind shore, clouds form and in many cases heavy snowfalls occur. Ice nuclei introduced into the clouds containing supercooled droplets should increase the number and decrease the size of ice crystals. The greater number of smaller particles, falling more slowly, should reach the ground farther inland than would normally be the case. This sequence of events has been modeled numerically, and the experimental program was based on the results of numerical model experiments. Results of the few field experiments conducted have been encouraging, but not conclusive. Field experiments were terminated in 1973 as part of the program of replacement and upgrading of the NOAA Research Flight Facility.

Cascade Snow Distribution Project. A similar effort, not designated as a national project, has been carried out by the University of Washington with Bureau of Reclamation support. The objective is to test the feasibility of diverting snowfall from the west slopes to the east slope of the Cascades. In this project more experiments have been initiated than in the Great Lakes program, but results are still inconclusive.

No project has yet been instituted to develop techniques for suppression of precipitation, but this would appear a reasonably extension of existing national projects.

Reduction of Severe Storm Winds

Hurricane Modification Project (Stormfury).   Attempts to modify hurricanes are based on the fact that release of latent heat by seeding results in expansion of the air in the cloud with consequent change of pressure. Under proper conditions, such variation of pressure can reduce the high winds that normally occur just outside the eyewall of the hurricane. In the first attempt at this technique, a mature hurricane was seeded in a try-it-and-see manner as it moved seaward 400 miles off the Florida coast in 1947. Following the seeding, the storm reversed course and ultimately caused extensive damage in Georgia. Whether the seeding was responsible for this disaster is unknown. In any case modification trials ceased abruptly, and were not attempted again until 14 years later.

Project Stormfury was initiated in 1960 by the U.S. Weather Bureau and the Navy with support by the Air Force, and has since been carried on during each hurricane season. Much of the effort has been directed at obtaining accurate observations of winds, cloud structure, rain, and other characteristics of hurricanes and also at developing numerical models for use in planning modification experiments. Under strict guidelines for selection of experimental cases, seeding has been limited to those hurricanes in which there is judged to be less than a 10 percent chance of the hurricane center's approaching within 50 miles of a populated land area within 18 hours after seeding. Few Atlantic hurricanes which are suitable in other respects have passed this test. Results from the first two storms seeded under these guidelines in 1961 and 1963 were encouraging to the scientists directing the program, but many more experiments are required before definitive conclusions can be drawn.[15] In 1969 Hurricane Debbie was seeded with results that could be interpreted as confirming a prior model calculation. However, serious uncertainties remained. Two years later Hurricane Ginger provided a fourth test, but the structure of this storm was so complex that it did not constitute a good prototype. The results were reported as favorable but not definitive. To obtain more complete data the Stormfury Advisory Committee and the NAS Committee on Atmospheric Sciences have recommended that field experiments be conducted in the Pacific Ocean as well as in the Atlantic. In early 1973 Project Stormfury was suspended pending acquisition of the instrumental aircraft needed for carrying on the program in the Pacific. Plans call for reactivation of Project Stormfury in 1976.

No national project has been designated with the objective of reducing winds associated with tornadoes, thunderstorms, or squall lines. Nevertheless, considerable effort is being channeled toward improved detection and measurements in such storms, and it is likely that if hurricane winds are successfully reduced, corresponding programs will be launched to reduce winds in other types of severe storms.

Reduction of Hail, Lightning, and Fog

Hail Research Experiment.   This project is carried on by the National Center for Atmospheric Research (NCAR) with National Science Foundation (NSF) funds and with aid from NOAA and several universities. Agricultural losses in the U.S. due to hail average about $300 million annually. The objective of this project is to establish a technique for increasing the number (thereby decreasing the size) of hailstones developed by severe thunderstorms. The field experiments of this five-year project began in 1972 in northeastern Colorado. Field experiments will be based on the results of numerical model calculations, and all pertinent characteristics of the storms will be measured using instrumented aircraft, radar, lidar, photography, infrared radiometers and scanners, and all other measurement techniques available. A major stimulus for this project has been the reports of remarkable successes achieved by an operational hail prevention program in the Soviet Union.

Lightning Suppression Project.   The Forest Service of the Department of Agriculture, which has carried on research in lightning suppression for more than ten years, has been designated as lead agency for this project. The objective is to determine how lightning strikes might be reduced in number or severity and to develop operational means for reducing lightning-caused forest fires. Analyses of field experiments in 1960, 1961, 1965, and 1967 indicate that some reduction in electrical activity may have been achieved. However, the results are not entirely clear, and the basic mechanisms are still in dispute.

Fog Modification Project.   The Federal Aviation Administration of the Department of Transportation, as lead agency for this effort, is aided by the National Science Foundation and Department of Defense. The objective is to develop effective operational methods for dissipating both cold fog (below 0°C) and warm fog (above 0°C) over airport runways and highways. Estimates indicate that for 41 selected airports, the cost of fog to airlines and ports in 1971 ran to $15 millions, while costs to passengers amounted to an additional $28 million.[16] The annual fog loss to airlines in the U.S. has been estimated at $75 million.[10] Programs are now in operation for dissipating cold fogs, therefore further work in this area is directed toward perfecting existing techniques and developing new ones. Warm fog, which accounts for about 95 percent of the total in the contiguous states of the U.S., is more difficult to dissipate because in this case the Bergeron-Findeisen process is not operative. Techniques have been tried that utilize heat to evaporate the fog droplets, helicopters have been used to pump air from a relatively warm dry layer down into the surface fog layer, and common salt and other hygroscopic materials have been used to encourage the selective growth of droplets. The first two of these techniques can be made effective, but they require large facilities and large expenditures of energy. Use of hygroscopic seeding agents presents the possibility of corrosion and is still of uncertain effectiveness.

Funding

Table 2 shows the total budgetary levels as tabulated by ICAS for fiscal years 1971, 1972, 1973, and 1974 for the general objectives of federal weather modification programs including the national projects. The numbers for FY1973 are estimated expenditures and those for FY1974 are proposed agency figures derived from ICAS Report No. 17[17] which was compiled prior to completion of Congressional authorizations and appropriations. The reported totals in a few cases do not represent exactly the tabulated figures, due both to round-off and to minor inconsistencies in the numbers reported by ICAS. The national projects constitute about half of the total federal effort directed to the modification of weather, exclusive of classified Defense Department expenditures.

In January 1973, half way through the fiscal year, FY1973 expenditures were cut back from the authorized level of $25.4 million to $20.2 million by direction of the Office of Management and Budget (OMB). At the same time the President's budget for FY1974 proposed a significant shift in the pattern of support for weather modification research. The Bureau of Reclamation program was reduced by about 50% while NOAA's budget was substantially increased, primarily for acquisition of instrumented aircraft. Subsequently, aircraft acquisition was deferred and support for the Great Lakes Project was suspended with the result that NOAA's proposed budget was reduced to below the FY1973 level. The total proposed for all agencies for weather modification research in FY1974 is 15% below estimated expenditures for FY1973. The decisions behind these changes have occurred at high administrative levels without direct participation by agency scientists or administrators. Whether these decisions represent an arbitrary reversal of laboriously developed policy or a temporary anomaly in the growth

Table 2
Agency Funding for Weather Modification
FY1971, 1972, 1973, and 1974
(Millions)

	Source: ICAS Nos. 16 and 17			
	1971	1972	1973	1974
Precipitation Augmentation and Redistribution				
Bureau Reclamation (Interior)	$ 5.9	$ 6.1	$ 5.8	$ 3.0
NOAA (Commerce)	1.6	1.3	1.4	1.3
NSF	.5[a]	.5[a]	.7[a]	.8[a]
	8.0	7.9	7.9	5.1
Reduction of Severe Storm Winds				
NOAA	.7	2.1	1.8	2.0
Navy	.1[b]	.1[b]	.0[b]	–
	.8	2.2	1.8	2.0
Hail Suppression				
NOAA	.5	.5	.5	–
NSF	2.1[a]	3.2[a]	3.6[a]	3.6[a]
	2.6	3.6	4.1	3.6
Fog Dispersal				
FAA (Transportation)	.6	.4	.4	.1
NASA	.1	.1	–	–
NSF	.8	.7	.8	.8
DOD	1.4	1.8	1.7	1.5
	2.9	3.0	2.9	2.4
Lightning Modification				
Forest Service (Agriculture)	.4	.4	.4	.3
NOAA	.2	.2	.2	–
NSF	.3	.3	.3	.3
DOD	–	–	.1	.1
	.9	.8	.9	.7
Inadvertent Modification				
NSF	.3	.5	.5	.7
Transportation	.0	.7	.7	1.3
NOAA	.3	.6	.6	.9
	.6	1.8	1.8	2.9
Social Impacts				
NSF	.2	.3	.5	.5
Bureau Reclamation	.6	.6	.6	.3
	.8	.9	1.1	.8
TOTAL	$16.4	$19.9	$20.4	$17.4

[a]The NSF figures include estimated support for mathematical modeling which ICAS lists separately as totaling $400,000 in FY1971, $520,000 in FY1972, $600,000 in FY1973, and $600,000 in FY1974.

[b]Aircraft operational costs ($1.5 million in FY1972) are not included.

of weather modification research is uncertain at this time.

Research support in FY1972 for universities and nonprofit research institutions amounted to about 30% of the total budget shown in Table 2, $3.8 million from the Bureau of Reclamation and $2.1 million from NSF. The budgetary changes for FY1974 imply severe reductions in university research and training in cloud physics and weather modification.

About $2.5 million of the federal funding in FY1972 supported work carried out by private commercial operators. In addition, private sector expenditures amounted to about $750,000 per year in the U.S., and operations in foreign countries by U.S. companies amounted to about $2,750,000 per year.* These figures are subject to greater uncertainty than the federal government figures, and other estimates have been larger.

Critique

The national projects, designed to secure new information in important areas of weather modification, represent those portions of the total federal effort which individual agencies most desire to push forward. In approving them as national projects, ICAS has not attempted to order priorities as to scientific or technical feasibility, potential contribution to society's needs, or cost. Consequently the level of effort applied to the various objectives is not directly related to society's needs. As Table 2 shows, for example, in the last four years hurricane modification research has been supported at less than the funding level for fog dispersal, despite the fact that hurricanes may impose 10 to 100 times greater economic losses—to say nothing of the loss of life. This anomaly reflects lack of coordinated planning, or, more explicitly, the independence of the supporting agencies and the fact that their appropriations are handled by separate Congressional committees.

The following items in Table 2 deserve special comment. First, operational costs of defense aircraft used in Project Stormfury, which amounted to more than $1 million in FY1972, are not included in ICAS tabulations or in Table 2, although approximately equal costs for NOAA are included. Second, classified weather modification activities are not included in Table 2, and, as indicated earlier, these may be

* Estimates of private-sector expenditures have been provided in personal communications by Eugene Bollay, Environmental Research Laboratories, NOAA, and Keith Brown, North American Weather Consultants.

substantial. Finally, the entries included under Inadvertent Modification are somewhat arbitrary. The Environmental Protection Agency (EPA) meteorology program, which amounted to $9.9 million in FY1973, is not listed by ICAS under Modification although it could be argued that at least part of it could be. Also, only small parts of the Climatic Impact Assessment Program (CIAP) of the Department of Transportation appear in Table 2 at levels of $.7 million (10% of the total) for FY1973 and $1.3 million (20% of the total) for FY1974, even though the goal of the complete CIAP program is to assess the effects of a hypothetical fleet of supersonic transports on global climate.

In several areas of weather modification, the failure of crucial field experiments to yield definitive results can be traced, in part at least, to submarginal funding. For example, prior to FY1972, Project Stormfury was funded at less than $1 million per year, far below the level needed to investigate the dynamics of hurricanes or the possibilities of modifying them. Thus, a whole decade passed with little progress in understanding what can be achieved in this important area.

In the area of precipitation augmentation the Bureau of Reclamation program has constituted a broad-based attack on a variety of specific problems. The fifteen projects under this program have been carried out in separate states by separate groups varying widely in scientific competence. This administrative mode has had two outstanding results. First, it has encouraged the growth of manpower in the weather modification field through development of research groups at several universities and the support of private cloud-seeding companies. Second, it has fostered political support in the states participating in the Skywater Program. It has not, however, contributed much to the capability to undertake large coordinated experimental field programs. Plans for FY1974 indicate that in the future the Bureau of Reclamation may limit its activities to the Colorado and High Plains Projects, and that it may no longer provide much support for university research or training.

The social impacts of weather modification have been studied by small groups under NSF grant or contract for the past seven years or so. Until the past few years these studies had been supported at such a low level that the difficult tasks of technological assessment of specific projects could not be attempted. Recently, however, such assessments of operational programs in hurricane modification and precipitation augmentation have been supported by NOAA, the National Water Commission, and the Bureau of Reclamation.[18, 19, 20] These are valuable first steps toward the comprehensive assessments which will be needed in the future, but it is doubtful that they yet provide the thorough and definitive analyses essential to sound policy decisions. The following fundamental questions concerning technological assessments remain: How critically dependent are the benefit-to-cost ratios on the overly simple estimates of wind suppression or precipitation increase which form the basis for these assessments? Can the crucial characteristics of a cloud or a weather system be determined prior to an operational decision? Has the full range of important consequences respecting disbenefits and liability been considered? Effective weather modification awaits the answers to these difficult queries.

III. Impacts of Weather Modification

> "The philosopher, the social scientist, the artist, the writer, the natural scientist—all are intellectual brothers under the skin."
>
> Glenn T. Seaborg

The two previous chapters served as an introduction to the topic of weather modification, focusing on its physical bases and the scientific or technological constraints within which it operates. Past efforts and continuing projects within the field were reviewed. In the following section, the larger impacts and implications of weather modification operations are examined in their several nontechnical dimensions—economic, social (or socioeconomic), legal, and environmental.

Research and field experiments on weather modification have been conducted in recent years by individuals, groups, and organizations. These efforts, however, have led to largely nonconclusive—at times conflicting—opinions about their effectiveness. Moreover, experiments on some of the crucial problems have been on too small a scale to provide definitive results.

Can the uncertainties that now characterize weather modification be removed in the future? Clearly, physical and scientific knowledge is more complete for some forms of modification than for others. Also, operational readiness varies greatly, depending on the degree of understanding as well as on the complexity of decision-making in a given situation. Despite present shortcomings, the prospects for successful

31

weather modification are promising enough that efforts will continue toward developing effective applications. It is therefore imperative that its associated benefits and disbenefits be weighed as carefully as possible in order to contribute to the development of a socially useful technology.

The Economic Aspects of Weather Modification

The economic evaluation of any public investment in natural resource development or management is rarely simple and straightforward. The now rich literature on benefit-cost analysis is testimony to the number and severity of some conceptual and a host of measurement problems. The very nature of the weather modification techniques that are presently or prospectively feasible in a technical sense makes it even more difficult to translate them into economic gains and losses. And, as most economists now stress, economic outcomes are only one part, albeit a major one, of the set of social results that must be weighed in deciding where, when, how, and on what scale a particular weather modification procedure warrants public investment.

A few of the major pitfalls that beset some current estimates of the economic effects of weather modification should be explored briefly. One is a common tendency to treat gross benefits as net. A 10 percent increase in precipitation in a given area may well increase its potential for crop production, but not without a variety of associated costs, since water is only one of the additional inputs required. Similarly, net reductions in fire damages that conceivably might be achieved through lightning suppression would be highly variable. For some types of timber, occasional small burns are actually beneficial to long-term productivity; in others they may be disastrous. In some cases, recovery of merchantable wood from a burned-over area may approach normal yields; under different environmental conditions, the entire output may be lost.

Another limitation is that net benefits available from a given set of weather modification procedures are bounded by the cost of the next-best way of achieving the desired objective. The major purpose of augmenting precipitation may be production of greater quantities of food and fiber. But weather modification is only one way of providing more water; moreover, a greater supply of water is only one way of increasing farm output and perhaps an inferior alternative to more and better capital equipment, greater genetic knowledge, or better developed human skill. In short, weather modification results must be valued in the context of the social objective to be sought. A benefit-cost ratio greater than one is a necessary but not a sufficient condition to establish the economic feasibility of weather modification; it must also be more efficient than any other way of realizing the same output.

This limitation is not confined to the case of water. Damages from violent storms and hail are all too real, and weather modification holds promise as a way of reducing them. But there are alternatives: mandatory storm insurance—alone or in conjunction with stiffer building codes to achieve storm-resistant structures—is an example. Another would be selective zoning to minimize the potential for damage in vulnerable areas. Indeed, two alternatives that appear attractive in many cases are to do nothing at all about modifying the weather but to adjust farming and other activities to its natural variations, and to invest in more accurate forecasting.

In brief, the essence of technological progress is the widening and deepening of man's ability to achieve given ends with the means at his disposal. This implies a need to expand the range of alternatives that must be weighed to assure a best solution. Given the inevitable budget constraints, some projects will always be feasible and others not.

A corollary difficulty may arise when costs are compared. For example, when water output from precipitation modification is valued at the cost of the next-best way of producing an equal amount of water, is there a demand for water at that price? If not, the cost of producing implies nothing as to its value. Is water from the alternative source available on a more predictable and/or controllable basis? If so, its economic contribution will obviously be greater than an equivalent annual increment that accrues irregularly and which is only partially predictable.

Finally, it should be noted that many of the important outputs from precipitation enhancement are not sold at market-determined prices, but at prices established by various subsidy devices. To the extent that the water produced is artificially valued as part of a scheme to transfer incomes from general taxpayers to farmers, a distorted picture is presented of the benefits of weather modification technologies.

In appraising the costs of achieving social benefits from weather modification, all costs of information must certainly be included. Precipitation management, for example, will produce economic benefits only if farmers take advantage of the increments by altering production schedules, processes, or cropping patterns. But this requires that two sets of costly and interrelated information be developed. The operating agency must convey to the decision-makers some level of information about both long-term yields (and variances) and current modification activities. And the users must develop their own capacity to utilize this information.

Weather modification projects are directed at one of the following two general objectives: to reduce losses from weather-caused dangers or disasters, or to increase the availability of a certain weather element, such as precipitation. Obviously, these will be perceived as good objectives by large parts of the public, so weather modification proposals start with a certain *a priori* presumption in their favor. As noted above, however, such perceptions may be oversimple and therefore deceptive. Rational decision-making demands that these benefits be quantified and weighed to the limit of present knowledge.

Estimates of benefits and costs have been made for certain proposed programs of weather modification. The Interdepartmental Committee for Atmospheric Sciences (ICAS) has reported estimates of benefit-cost ratios of from 2 to 1 up to 20 to 1 as likely to result from current pilot projects.[10] Potential benefits and costs of some of the national projects referred to in Section II are briefly reviewed below. These benefits, as well as the estimates given above, should be accepted only with the limitations cited earlier. In most cases, they cannot be regarded as conclusive.

Precipitation Augmentation
and Redistribution

The Colorado River Basin Project is designed to augment seasonal precipitation and snowpack through cloud seeding. According to a report of the Interdepartmental Committee on Atmospheric Sciences, irrigation benefits alone (for areas served by the basin) are estimated at $50 per acre-foot/year. The increase in runoff resulting from the snowpack augmentation, based on an assumed increase of 15 percent in total snowpack due to seeding, is conservatively estimated to be two million acre-feet, at a cost of about $1.50 per acre-foot. To these should be added other economic benefits due to increased hydroelectric power and salinity control.[10]

In the Connecticut River Basin, estimates of 15 percent precipitation augmentation from a managed atmospheric water resource system would result in an incremental runoff of 2 million acre-feet of water.[10]

In areas of Florida, traversed by moisture-laden but sometimes non-precipitating or lightly precipitating cumulus clouds, cloud seeding has been used in both research and quasi-operational programs. During periods of drought, such augmentation can be worth $50 per acre-foot, whereas experiments have indicated costs of $1 per acre-foot in certain areas.[10]

S 2003170

Modification of Hurricanes and Severe Storms

Among all atmospheric phenomena, hurricanes are responsible for the most extensive property damage and fatalities. Hurricane damage in the United States averages about $500 million per year, and the losses in property when major hurricanes strike over land may exceed $1 billion for each occurrence. Although accurate assessment is nearly impossible, available data on property damage suggest that a reduction of 15 percent in maximum winds could possibly reduce property losses by 50 percent.[18] Such reduction would also be expected to save lives. Results from Project Stormfury (described earlier) indicate a hope still unproven for modest reduction in wind speeds. The ICAS report cited earlier states that a reduction of only 10 percent in losses would provide benefit-cost ratios in excess of 10 to 1.[10] For a single hurricane such as Betsy (1965) or Camille (1969), it has been suggested that seeding might reduce property damage by over $200 million. Camille in 1969 ravaged several states, causing damages amounting to $1.4 billion and almost 300 fatalities.[6, 18] In June 1972 flooding associated with Hurricane Agnes resulted in 118 deaths and in property damage estimated at $3.5 billion, the largest economic loss from a single natural disaster in the nation's history.

Tornadoes, although small and short-lived, represent a serious threat to people and property. On the average, it has been estimated that in the United States they cause 125 deaths and $75 million of damages per year. Tornadoes occur frequently, but only the largest ones pose serious threats—it appears that 1.5 percent of the storms may cause 85 percent of the fatalities. Some tornadoes, however, occur in association with hurricanes, thus the exact damage they cause may not be readily identified. The detailed structure and behavior of tornadoes is not fully understood; the storms are small, of brief lifetime, and scattered geographically. For these reasons, it will be difficult to develop effective means for modifying tornadoes.[6]

Fog, Hail, and Lightning

Efforts to dissipate supercooled fog (droplets with temperature below 0°C) have been economically successful. Such efforts by U.S. airlines at several airports resulted in benefit-cost ratios of more than 5 to 1 in savings and in preventing delays and diversions of traffic.[10]

Warm fog dispersal is still in an exploratory stage. Since 95 percent of all fogs occurring at U.S. airports (excluding Alaska) are warm, the problem is of considerable importance. According to the Airline Transport Association estimates, delays caused by these fogs account for annual losses of about $75 million for U.S. airlines. One occurrence may cause as much as $100,000 of revenue loss to airport users at a major airport, and even higher figures must be expected in the future as very large planes come into wider use. Highways and harbors also suffer from fog-caused transportation delays, damages, injuries, and deaths.[6, 10]

Although reduction of hail damage has been reported in some cases where thunderstorm cells have been seeded, these results should be regarded as experimental. However, annual hail damage in the U.S. amounts to more than $300 million, and commercial seeding operations in limited areas to protect certain crops have reportedly resulted in benefit-cost ratios of better than 2 to 1.[10] It is notable that in the Soviet Union hail suppression operations have been underway for several years, and the operators claim that the value of crops saved has exceeded seeding costs by a factor of 10 or more.

Lightning is considered to be the greatest single cause of forest fires in the western U.S., burning over two million acres annually. To this loss is to be added the costs of human lives and of an estimated $100 million fighting the fires. Experiments conducted as part of the U.S. Forest Service Project Skyfire expose the complexity of the phenomenon. Artificial seeding, although it may decrease lightning activity, can also affect the behavior of the cloud in other ways. Release of latent heat in certain instances may significantly increase the strength of vertical convection and increase the cloud's electrical activity. Positive results have been reported from specific experiments such as those carried out during the summers of 1960, 1961, 1965, and 1967.[6, 10]

This summary of benefits and costs relates to possible applications of results from the work being done under certain of the national projects. The data presently available suggest that economic benefit-to-cost ratios for some weather modification projects are likely to be positive and large enough to encourage further research and development. However, more accurate data than now exist are needed to assure that any given weather operation will be both feasible and economically worthwhile.

Such information, however, represents only one facet of the problem. The other side of the coin is represented by the "externalities"* or external effects of weather modification activities. Some of these externalities are solely economic, but others extend beyond the realm of economics to exert impacts that are social (and socioeconomic), legal (and political), and environmental/ecological.

These external effects are difficult to identify and even more difficult—in some cases, impossible—to quantify or measure. The effects of weather on human activities are varied, ubiquitous, and undocumented, and little is known of the relationships between weather and many industries. And still less knowledge is available on how weather changes will affect the overall economy.[21]

In certain areas of the economy weather modification may be expected to have fairly direct and simple consequences. Such areas may include the following.[21, 22]

Agriculture:	increased precipitation at the right time and place, hail suppression,
Hydrosystems:	increased precipitation and/or snowpack,
Fisheries:	increased precipitation to augment river flow,
Aviation and Transportation:	fog dispersal, severe storm modification.

Other operational programs may initiate unintended effects of complex. and sobering proportions. Hurricane modification and regional changes in precipitation fall in this category. In these cases, both benefits and disbenefits may be distributed far beyond those directly concerned with decisions to modify or not to modify.

Social or Socioeconomic Aspects of Weather Modification

Among the externalities that accompany weather modification efforts, the social implications are perhaps the most elusive and difficult to evaluate since they impinge on the vast and complex area of human

---

*An "externality" is an unpriced effect. It may be a benefit received by those who do not pay for it or a loss incurred by those who are not compensated.

values and attitudes. In examining this problem of social effects, Sewell has suggested the following questions: Should we modify weather at all? If so, when, where, and how far should we go?[21] However, these may not be the most useful questions to ask. It may be more important to arrive at certain guidelines and procedures which should govern the decision-making process.

In regard to the social implication of any weather modification project, the following important elements go beyond the purely economic evaluation.

1.   The individuals and groups to be affected, positively or negatively, by the project must be defined. An operation beneficial to one party may actually harm another. Or an aggrieved party may hold the operation responsible, say, for damage to his crops which might occur at the same time or following the modification. This problem has clear legal implications, as will be shown.

2.   The impact of a contemplated weather modification effort on the general well–being of society and the environment as a whole must be evaluated. Consideration should be given to conservationists, outdoor societies, and other citizens and groups representing various interests (or no interest other than a feeling of responsibility and of duty as citizens) who presently tend to question any policies aimed at changes in the physical environment. It is reasonable and prudent to assume that, as weather modification operations expand, questioning and opposition by the public will become more vocal. A Pennsylvania case, *Pennsylvania Natural Weather Association v. Blue Ridge Weather Modification Assoc.* (1968) demonstrates this public concern. In this case, the plaintiff claimed that activities of the defendants were wrongful on counts generally pertaining to the public interest, i.e., releasing of chemicals in the air dangerous to the health, creation of a nuisance, trespassing and unreasonable use of a natural resource, etc.

3.   Consideration must be given to the general mode of human behavior in response to innovation. Attempts made by the Quebec Hydro, beginning in 1964, to raise the levels of its reservoirs by inducing increased precipitation provide a case in point.[21] Earlier attempts had been unopposed. In this case, however, cloud seeding was followed by precipitation which continued for three weeks, exceeding all expectations, and local residents who suffered resultant economic losses not only protested at the time but continued to protest, and even to threaten violence, after all operations had been suspended.

4. The uniqueness and complexity of certain weather modification operations must be acknowledged, and special attention should be given to their social and legal implications. The cases of hurricanes and tornadoes are especially pertinent. Alteration of a few degrees in the path of a hurricane may result in its missing a certain area (heavily populated, perhaps) and ravaging, instead, a different one. The decision on whether such an operation is justified can reasonably be made only at the highest level within the executive branch of the government, and would need to be based on the substantial scientific finding that the "anticipated" damages will be less than those originally predicted had the hurricane been allowed to follow its course. The legal implications in this area are, again, sobering.

Complexity surrounds the hurricane phenomenon—in its physical aspects, in the making of decisions regarding its modification, and in the ramifications of any such operation. The modification of hurricanes conceivably may result in aftereffects greater than the intended direct effects, as indeed may some other future modification projects. On the other hand, some weather operations are relatively straightforward with respect to both technique and effect. Fog dispersal in airports, for example, with its clearly defined advantages, presents a simple problem, where harmful effects are unlikely. If lightning could be suppressed without decreasing precipitation, this would constitute another case of clear-cut benefits.

5. Attention must be given to alternatives in considering a given weather modification proposal. The public may prefer some other solution to an attempt at weather-tampering which may be regarded as unpredictable and risky. Furthermore, alternative policies may tend to be comfortable extensions of existing policies, or improvements on them, thus avoiding the public suspicion of innovation. In an area such as weather modification, where so many uncertainties exist, and where the determination or assigning of liability and responsibility are far from having been perfected, public opposition will surely be aroused. In agriculture, for example, the following alternatives should be explored: other water sources, other land use (particularly in the case of arid areas), other products and methods of production, and improved ability to control the seasonal distribution of rainfall. Any alternative plan or combination of plans will have its own social effects. An alternative agricultural product, for example, may require new working habits in terms of time spent in the fields, harvest time, and labor needed; and it may result in a change in productivity and in net income.

It is the overall impact of an alternative plan and the adverse effects of not carrying out such a plan which, in the final analysis, should guide decisions on alternative action.

6.   Finally, it is important to recognize that the benefits from a weather modification program may depend upon the ability and readiness of individuals or groups to change their modes of activity. The history of agricultural extension work in the U.S. suggests that this can be done successfully, but only with some time lag and at a substantial cost. The work of Sewell, Kates, and Phillips[23] and others suggests that public perception of flood, earthquake, and storm hazards is astonishingly casual. And Sims and Baumann have suggested that marked geographical differences in death rates due to tornadoes may be associated with differences in psychological attitudes toward uncontrollable natural events.[24]

Even when a full range of probabilities is made available and explained, the inherent complexity of the link between weather events and the individual householder or business unit is so weak that correct action does not seem to follow. This problem would seem likely to expand as man's capacity to modify weather increases. It is not immediately obvious that those who share benefits and costs will know it, understand it, care about it, and act appropriately. At the very least, a large-scale, continuing program of education (and perhaps some compulsion) will be required if the potential social gains from weather modification are to be realized in fact.

Legal Implications and Impacts of Weather Modification

Because of the novelty and the relatively recent origins of weather modification, traditional legal concepts may not be expected to substantially govern its present practices. Despite interest shown in the subject in government circles, little federal legislation has been enacted, and in that the emphasis has been almost entirely on research and evaluation. On the state level, twenty-nine states have enacted laws in the field, of varying scope and context. Only eight cases involving weather modification effects have been handled by the courts to date.[25] The historical development of federal and state statutes pertaining to weather modification in the last twenty years is summarized here, followed by a discussion of the third legal aspect of weather modification—namely, private litigation.

Federal Statutes

The first bill aimed at regulation in the field of weather modification, the Weather Control Act S. 4236, was introduced in 1951. It would have created a Weather Commission with the objective of developing a program of federal control of experiments and operations; however, this act died after it was referred to the Senate Commerce Committee. In 1953, an Advisory Committee on Weather Control was created by P.L. 83-256 and was charged with studying and reporting on the need for regulation. The Committee's report of 1957 acknowledged the potential of weather modification and recommended further study and research by the NSF. On the legal side, it "recognized that neither the meager case law, nor the few state statutes then in effect, could 'prevent cloud seeding under the wrong meteorological conditions or by professionally unqualified or poorly equipped operators'."[25]

As a result of this recommendation Congress authorized the NSF (by P.L. 85-510 of 1958) to initiate and support a program of study, research, and evaluation in the field of weather modification. The NSF was also to encourage voluntary filing of information by weather modifiers on their activities, and it later undertook the coordination of federal agency programs by holding annual interagency conferences on weather modification.

From 1966 to the present, efforts toward federal legislation have increased. This period began with the introduction of two competing bills representing the views of the two principal agencies seeking lead status in the field of weather modification—the departments of Commerce and Interior. These bills, and most other relevant legislation of the past five years, have been discussed in Chapter I of this study. Some of the federal efforts which have not resulted in legislation may be mentioned here briefly. An extreme proposal, H.R. 8977, sought to prohibit weather modification anywhere in the nation. In 1969, S. 1182 recommended creation of a nine-member commission to be appointed by the President to study the need for federal regulation. Finally, S. 3919, introduced in 1970, authorized the Environmental Science Services Administration (ESSA, the predecessor of the National Oceanic and Atmospheric Administration [NOAA] in the Department of Commerce) to establish a mandatory reporting system for all weather modification activities in the U.S. This, however, was designed as an interim bill to fill a gap until more comprehensive action could be taken by Congress.

State Statutes

Lack of uniformity, and even contradiction, characterize the weather modification regulations thus far passed by twenty-nine states. A few examples may illustrate these varying attitudes.

The states of California, Pennsylvania, Texas, and Washington have enacted comprehensive legislation allowing weather modification only after compliance by the modifier with licensing, project permit, and notice requirements. On the other hand, Connecticut, Hawaii, New Hampshire, and Oklahoma have adopted statutes simply noting the interest of their legislatures in the subject. A few others, such as Idaho, Utah, and Wisconsin require only that weather modification operations be registered with the proper state agency.

Texas has officially declared by legislation that weather modification is not inherently ultrahazardous or subject to absolute liability rules in private damage action. In contrast, Pennsylvania and West Virginia have adopted laws imposing absolute liability in private lawsuits arising from weather modification. Maryland took a step further by banning all weather modification.

This diversity in attitudes reflects the states' widely varied experiences with weather modification, as well as the differences in attitudes toward exploitation of natural resources referred to in Chapter I. Johnson[25] notes two broad patterns or classes: those states providing for active control of weather modification, including the collection and evaluation of scientific information; and those concerned primarily with the latter and imposing only minimal licensing requirements. In states composing the first and larger group, the licensing and registering of operators is assigned to administrative agencies. Licenses are subject to requirements covering qualification of modifiers, financial responsibility, and the nature of the proposed project.

Finally, some states, including Pennsylvania, West Virginia, Colorado, and New Mexico, have passed laws dealing specifically with weather modification that might have effects outside their own borders. Others, such as Louisiana, Nebraska, New Mexico, North Dakota, South Dakota, and Wyoming, claim sovereignty over the moisture in the atmosphere.

Private Litigation

All cases argued by the courts thus far have concerned damages allegedly resulting or anticipated from cloud seeding operations. These cases are reviewed in this section, and then some of the problems are discussed which can be anticipated when operations with far-reaching socioeconomic and legal implications may become established practices.

The present legal system is inadequate to cope with problems of weather modification. This is particularly true in questions of liability and indemnification for damages believed to have been caused by cloud-seeding activities. As demonstrated by the few cases already adjudicated, the plaintiffs have been unable to establish cause–effect relationships, and this inability emerges both as the basic issue and as the cause of the legal incapacity. For the plaintiff to satisfy the burden of proof of causation is a nearly impossible task, not only because of the novelty of the legal question, but also because of the complexity of cloud processes and the difficulty of making complete and accurate measurements of the many variable quantities.

It is not surprising that no significant body of common law has yet emerged. One would expect, of course, that as more cases are examined, such a body of common law will slowly emerge, simplifying the litigation issue. It should be hoped, nevertheless, that cases stemming from the rather sensitive and highly controversial institution of weather modification will, to some extent—and within the boundary of legal common sense—continue to be examined on their respective individual merits. The social and human implications of any given case may be quite unamenable to statutory classifications. This is not, however, to undermine the significance of legislation—for which there is real and immediate need—as the basic juridical source for dealing with litigations resulting from weather modification operations.

Tort ("wrong") law—at least from a theoretical standpoint—provides by analogy a source of application to weather modification activities. Two elementary questions particularly relevant to this and similar problems are: Who owns the clouds? On what basis is ownership said to exist? As already noted, some states have answered this question by asserting sovereignty to the moisture in the atmosphere.

No attempt will be made here to discuss in detail the various liability theories such as tresspass, negligence, strict liability, and nuisance,[27] or the other legal analogies drawn from various laws and now associated with tort law, which include rule of capture, law of riparian rights, doctrine of prior appropriation, the natural right theory, and the ownership of air space theory. Instead of a discussion of such theoretical or abstract concepts of laws and analogies, it seems more useful here to review the actual cases involving weather modification thus far decided by courts. Certain cases may be seen to relate, although not always in clear terms, to one or another of these concepts.

Since 1950, eight lawsuits involving weather modification have been decided. In most of these cases, court decisions have been in favor of weather modifiers. Only in two (actually three, since two cases were decided as one) decisions were temporary injunctions or restraint orders issued against modifiers.[25, 26, 28]

The first group of cases includes the following:

Slutsky v. City of New York (New York, 1950). This is the only case where cloud seeding was held to be permissible even if the objecting party (a resort owner) might be injured. The plaintiff's petition for an injunction was denied and the court held that the need of the ten million inhabitants of New York for water was real, whereas possible damage to the plaintiff's property was speculative.

Samples v. Irving P. Krick, Inc. (Oklahoma, 1954). This case was decided in favor of the weather modifier on the basis that the plaintiff could not prove to the satisfaction of the jury that property damages incurred were caused by the defendant's cloud-seeding operations. The plaintiff in this case advanced charges of negligence and carelessness on the part of defendant and hence held him liable.

Adams et al. v. the State of California (California, 1964). This case was decided in favor of the weather modifier on the same basis as in the preceding case. The court was not convinced that defendant's modification attempts were the cause of a flood which damaged plaintiff's property.

Pennsylvania Natural Weather Association v. Blue Ridge Weather Modification Association (Pennsylvania, 1968). In deciding this case, the court refused to enjoin the defendants from carrying out a hail suppression program because there was no proof that plaintiffs were deprived of any water as a consequence. As pointed out earlier, charges by plaintiffs in this case referred specifically to particulars such as danger to health, nuisance, trespass, etc. It is of interest here to note that in the other Pennsylvania case, *Township of Ayr v. Fulk* (1968), where the court decision upheld a township ordinance prohibiting weather modification, potential pollution was cited as one of the dangers posed by cloud-seeding.

The second group of cases, those where temporary injunctions against modifiers were issued, comprises the following:

Auvil Orchard Co. v. Weather Modification, Inc. (Washington, 1956). A temporary restraining order was granted the plaintiff by the court against defendant's hail suppression attempts on the ground that they caused crop damage. After hearing expert testimony, the court refused to make the injunction permanent because it was not convinced that the exceptional rainfall or floods were caused by cloud seeding. In this case it is clear that the temporary restraining order effectively stopped defendant's modification attempts for the season.

Southwest Weather Research, Inc. v. Duncan; and Southwest Weather Research, Inc. v. Rounsaville (Texas, 1958). The court temporarily enjoined modifiers from seeding clouds for hail suppression until it could be shown that their activities would not reduce the amount of precipitation on the plaintiff's lands. The temporary injunction became, in effect, a permanent one, as no evidence was subsequently introduced by defendants. One may note, however, that even if the temporary injunction had been lifted, the modifiers had lost at least one season of operation.

The question of modification of hurricanes and tornadoes is another area where legal problems of tremendous magnitude may be involved, in addition to the social implications mentioned earlier. Because of the menace of each single hurricane in terms of human and economic losses, some questions pertaining to the legal aspects must be faced.

1. Who should have the power (an individual, or agency, etc.) to authorize modification in the path of a hurricane—particularly, if it is possible to determine, *a priori*, that the new course will result in damages and loss of lives in areas which would have remained safe had the hurricane been allowed to follow its original course?

2. How, and by whom, can those affected by the "modified" hurricane be compensated for their losses? And what about those who lose their lives? Should those who benefit from successful modification be assessed?

3. How, and by whom, will those who have been deprived of certain amounts of rainfall due to the change in the course of a hurricane be compensated? Rain produced by hurricanes contributes significantly to the water supply for large areas in the southern and eastern U.S.

4. Who will be responsible for the consequences of the operational and design aspects of a project and for its efficient execution? What if the results differ from those foreseen by the decision-maker?

5. Who will be accountable for the consequences of "no modification"? In other words, if the decision is to avoid any action for fear of liability or for whatever other reason, who is responsible for the consequences? The Port of New York Authority has recently suggested that legal redress may be sought under either condition, i.e., modification or failure to modify.

6. How will insurance companies react to a situation where modification of a hurricane has caused damage and possible loss of life in an otherwise predictably safe area?

7. How will the U.S. government respond to a situation in which a seeded hurricane may cause loss of life, damage to property, or reduction in rainfall in another country?

The scope of these questions emphasizes the need for new legislation on the federal level to establish a basis for resolving interstate and international issues. This need will become urgent if activities in this field should move from research into large-scale operations. Also, the interrelationships between legislation and scientific developments should be kept in mind. The legislation must facilitate, not hamper, the use of new information in the development of policy as research results warrant.

Environmental Impacts of Weather Modification

Among the various aspects of weather modification activities, the least defined are those related to environmental and ecological impacts. Little information is available on this subject despite the acutely sharpened public awareness of activities suspected of having potential harmful effects on the environment.

This lack is understandable in the case of deliberate weather modification because of its fairly recent introduction. On the other hand, inadvertent weather modification is not a new phenomenon, but one which has always accompanied urbanization and industrialization. Especially in recent decades the increased consumption of fossil fuels has added great quantities of carbon dioxide and other gases and particles to the atmosphere. Oceans and other bodies of water are increasingly contaminated by the dumping of oil and other chemical effluents. These changes undoubtedly degrade the earth's ecosystems and subsystems, but quantitative date are remarkably imprecise.

The difficulty of assessing the impact of weather modification is compounded because changes are imposed on an already variable climate where "normal" weather changes from one year to the next. Also, species populations of many plants and animals fluctuate on a variety of sites, adding further to the problem of determining which changes are due to weather modification.[19]

In view of these complexities it is perhaps not surprising that the environmental impact statements which are required under Section 102 of the National Environmental Policy Act of 1969 (P.L. 91-190) have had little effect on weather modification planning or programs. The required impact statements are each expected to discuss the overall impact of a particular project, any adverse effects which are foreseen, and the possibility of long-term effects. Among other matters to be included are an evaluation of alternative courses of action and consideration of possible irreversible commitments of resources.

By the end of 1972, impact statements had been filed for eleven weather modification projects; five of these had been accepted by the Council on Environmental Quality (CEQ) as final statements and six were in draft form. In reviewing the available statements one cannot avoid the impression that for the most part they represent *pro forma* responses to the law, rather than thorough professional analyses of the complex environmental effects which they purport to discuss. For example, the impact statement filed by NOAA in support of Project Stormfury is limited to brief reassurances that direct effects of seeding hurricanes will be benign, and no attention is directed to some of the serious policy issues associated with such efforts. However, the requirement of impact statements is still new; it constitutes potentially the most powerful mechanism now available to the federal government and to the public for insuring that weather modification activities are carried out in the public interest.

Environmental concerns also have been expressed at the state level. For example, a 1972 referendum held in the San Luis Valley of Colorado rejected further modification of weather and natural precipitation by a ratio of 4 to 1.[29] Subsequently, a hearing before the Natural Resources Director denied an operating permit on the grounds that proof of economic benefit was lacking.

Recent years have brought a significant change in the public attitude toward problems of the environment which may lead to increased research and to more responsible application of acquired knowledge.

Man will have to manage his environment. But to do so most efficiently and with the smallest possible detrimental impact, he must use great wisdom in determining priorities and courses of action.

International Aspects

Because the atmosphere cannot be confined within national borders, efforts to modify weather may have international ramifications. Some of these have been referred to in earlier sections: downwind effects of cloud seeding, possible changes in the path or severity of hurricanes, increased acidity of rainfall by air pollution, and the use of cloud seeding for hostile military purposes. In each case the regional and national impacts of weather modification are extended to the international arena, in some cases with consequences which are potentially more serious.

Clearly, to experiment with hurricanes which might pass over another country is to invite a variety of international problems and difficulties. At the same time, the ability of the United States to modify hurricanes in a beneficial manner could contribute significantly to international good will and understanding, especially in the Caribbean, Central America, the eastern Pacific, the Bay of Bengal, and Japan. And any capability which might be developed for alleviating drought would be of immense benefit to those areas of the world subjected to these disastrous weather anomalies. These are substantial reasons for the federal government to initiate international discussions directed toward agreement on objectives, procedures, and safeguards before field experiments are conducted which might affect other countries.

Military use of weather modification techniques also may constitute an international issue of serious dimensions. The scientific community has expressed strong and united opposition to such activities. Three primary reasons underlie this opposition. First, it is recognized that free and immediate exchange of weather data is essential to the existing worldwide system of weather observing and forecasting. Furthermore, the nations of the world have undertaken in the past few years a highly coordinated long-term research program addressed to extending the range of forecasting and to understanding changes of climate; this program depends critically on the high level of international cooperation and the open communications which have long characterized this field. Obviously, hostile use of weather modification would seriously jeopardize this vital international cooperation. Second, as stated earlier, further scientific progress in weather modification requires carefully planned and executed field research programs. Operations carried out on a large scale, especially operations carried on secretly, could not meet these standards and would run the risk of contaminating the results of experimental programs. Third, many scientists recognize military weather modification as a first step toward other possibly more terrible forms of geophysical warfare, such as initiating landslides or earthquakes, a step they believe should not be taken.

Resolutions concerning this issue have been introduced into the Senate and the House of Representatives, and hearings were held on July 27, 1972 by the Senate Foreign Relations Sub-committee on the Ocean and International Environment. In this connection the Defense Department has consistently refused to provide information to Congress concerning its weather modification operations in Southeast Asia.[30] On June 25, 1973 Senate Resolution 71 (introduced by Senator Claiborne Pell) was passed by a vote of 82 to 10 urging the U.S. government to seek an international agreement to ban weather or other geophysical modification as a weapon of war. Thus, the issue has been joined between Congress and the Defense Department. Clearly, the international dimensions of this issue require that it be evaluated and resolved at a level above the Defense Department.

In addition to these examples, inadvertent weather modification also may have important international ramifications. Although this subject is poorly understood, it is a matter of increasing concern, as is evidenced by the United Nations Conference on the Human Environment held in Stockholm in 1972. Future expansion of industrialization and changes in land use over large areas of the earth will be likely to result in greater attention being directed to possible inadvertent effects. In response to this concern, the NAS Committee on Atmospheric Sciences has urged establishment of an advisory mechanism for consideration of weather modification problems of potential international concern before they reach critical levels.[31]

IV. Weather Modification and the Requirements for Decision-Making

"It becomes one, while exempt from woes, to look to the dangers."

Socrates

Thus far in this report potential applications of weather modification have been identified, past and on-going projects have been described, and the benefits, disbenefits, and impacts of each have been discussed. In this section the requirements for making rational decisions in the field of weather modification will be addressed explicitly, both as they relate to research programs and to future operational projects.

The process of decision-making is of necessity strongly influenced by the uncertainties associated with the particular weather modification proposal. For example, the approach to a decision will be far different for operational dispersal of cold fog than for experimental modification of hurricane winds. In fact, a vital part of decision-making should be to rate the proposed activity on the scale of uncertainties. The following discussion considers various important elements required for decision-making, but it does not extend to prescribing the process to be followed in any specific case.

Scientific and Technical Information

The primary requirement for reaching sound decisions in weather modification is the provision of adequate and reliable information. Three critical aspects of this problem can be discussed under these general headings: research coordination and leadership, national laboratory, and project assessments.

Research Coordination and Leadership

The need for strengthened research in weather modification has been pointed out earlier in this report. To define the conditions under which cloud seeding will increase or decrease precipitation by specific amounts, to determine whether and how hurricane forces can be reduced, or to determine the effects of men's activities on climate, substantial, sustained programs of theoretical, laboratory, and field research will be required. Previous efforts along these lines have been submarginal in manpower and in the facilities available, and programs have been largely fragmentary and uncoordinated. What clearly is needed is the assignment of responsibility for crucial research efforts to a single government agency, together with provision of adequate resources. This *lead agency* should take the initiative in planning and carrying out research on key scientific and technical problems and should serve as the central focus for the national program in weather modification. This does not mean that all research or operational programs should be collected into that single agency; the missions of several federal agencies require that they be involved in such research. With regard to support of nonfederal research and development by universities, nonprofit research institutions, and industry, the system of diversified support that operates in other research fields has distinct advantages in helping to insure that research is as broad in perspective and unbiased as possible.

In 1971 the Interdepartmental Committee for Atmospheric Sciences (ICAS) designated lead agencies for the following seven national projects described in Chapter II of this report.

National Colorado River Basin Pilot Project—Bureau of Reclamation
National Hurricane Modification Project—National Oceanic and Atmospheric Administration (NOAA)
National Lightning Suppression Project—Forest Service
National Cumulus Modification Project—NOAA
National Hail Research Experiment—National Science Foundation (NCAR)
National Great Lakes Snow Redistribution Project—NOAA
National Fog Modification Project—Federal Aviation Administration

Thus a central focus of responsibility has been established for each of the national projects, but no single lead agency is accountable for overall leadership or coordination. Allocation of resources to the various national projects remains under separate federal departments or agencies and under separate Congressional committees. The need for a single lead agency becomes more acute as the national weather modification program increases in size and complexity.

The NAS Committee on Atmospheric Sciences has recommended in *The Atmospheric Sciences and Man's Needs: Priorities for the Future* that NOAA be assigned lead agency (but not single agency) administrative responsibility for developing a research program to attack the crucial scientific problems not handled adequately under on-going programs and for coordinating mission-oriented research of all agencies to enhance the results of the overall program. [31] This recommendation has been reiterated in *Weather and Climate Modification: Problems and Progress.* [6]

The Role of a National Laboratory

Certain of the crucial research programs needed to develop weather modification capabilities require the facilities of a major national laboratory. *The Atmospheric Sciences and Man's Needs: Priorities for the Future* listed the following such programs: "investigation of the dynamic effects that can be produced by cloud seeding in order to determine whether hurricane winds can be reduced in severity and their paths changed"; "the application of cloud-seeding techniques to modification of squall lines, thunderstorms and other severe storms"; "investigation of the atmospheric characteristics that distinguish positive effects of cloud seeding from negative"; and "determination of the optical properties of atmospheric particles in order to determine whether increased global concentrations will produce cooling or warming." [31]

Establishing the laboratory under the lead agency would facilitate planning and management and provide the necessary resources. If the laboratory were to be affiliated with another agency active in weather modification, its relationship with the lead agency would have to be defined carefully, and a maximum measure of cooperation and exchange of information assured between the two agencies.

A national laboratory should be expected to perform critically needed assessments of the technical, economic, legal, and social aspects of weather modification. It would bring to bear on the task the skills of atmospheric scientists, social scientists, statisticians, lawyers, and political scientists.

At present, the National Center for Atmospheric Research (NCAR), which is operated by a consortium of universities with NSF support, comes closest to being a national laboratory in the sense referred to here. Its primary role has been to develop research programs and to coordinate university studies concerned with problems too large to be handled by individual institutions. NCAR serves as lead agency for the National Hail Research Experiment, and it would be possible to build a national weather modification laboratory from this initial base. A national laboratory could also be developed within the NOAA Environmental Research Laboratories, or through redirection of an existing laboratory, or through a new university-affiliated laboratory supported by one of the agencies active in weather modification. Any one of these alternatives would be satisfactory but would depend upon availability of adequate funding and administrative leadership.

Assessments of Weather
Modification Projects

An important forerunner to decision-making in this emerging technology must be the provision of broad assessments of the effects of possible weather modification activities. Such studies not only enhance our understanding of the technical and social issues at stake, they also aid specifically in the process of decision-making by documenting more clearly the various elements which must enter into each decision. Three such assessments have been completed in the past three years—two of them concern precipitation modification, the third deals with decision analysis as applied to hurricane modification. A brief discussion of these studies follows.

Precipitation Modification. This study, published by the National Water Commission in 1971, had the expressed objective "to provide background for the Commission's deliberations on the subject of national water policy."[19] It also aimed "to stimulate public discussion of water resource policy issues." Attempting to discover whether precipitation modification can reliably and usefully augment the nation's water supplies, the Water Commission approached that question within the context of its responsibility "to develop policy recommendations about the physical aspects of water supply," and no attempt was made to discuss technical matters concerned with research programs or operations. The report reflected a review of the literature on precipitation modification and the views of a group of experts on the subject. These views, and the Commission's own posture on the subject, were represented in the conclusions and recommendations. The study included an appraisal of precipitation modification (orographic, convective, and cyclonic storms), downwind effects on usable water supply, social, economic, and ecological effects, and, finally, relevant legal and institutional problems. The report recommends concentrating research on the increasing of run-off during drought periods, because increases here would be of greatest value and importance.

Decision Analysis of Hurricane Modification.   This study was carried out by the Stanford Research Institute (SRI), and the report[18] demonstrates the breadth of considerations to be expected in certain weather modification operations. A summary has been published in *Science.*[32] As its initial objective, the study aimed "to identify ways in which decision analysis can make significant contributions to NOAA, both in its technical operations and in its management and planning function." However, as a result of discussions between the SRI, the National Hurricane Research Laboratory (NHRL), and the Department of Commerce, the focus of the study became the allocation of national resources to modification of tropical hurricanes.

An overview of the decision problems inherent in hurricane modification—the initial phase of the project—was used to guide the detailed analysis, which sought to incorporate the overall physical, economic, social, and legal dimensions defining modification operations. To help further in understanding the problem of assessments, both meteorological and economic models were developed; the former dealt with the incidence of hurricanes and with the effect of seeding on maximum winds, and the latter with property damage inflicted by maximum winds. The study recommended research "on a greatly expanded scale to provide a more refined basis for operational seeding decisions" and urged that "seeding should be considered on an emergency basis if a severe hurricane threatens the coast of the United States."

Assessment of Colorado Basin Project.   This report[20] of a Stanford Research Institute (SRI) study, carried out by an interdisciplinary team from SRI and the University of California at Davis and sponsored by the National Science Foundation, constitutes the most extensive assessment of any weather modification program. It is directed at determining the probable effectiveness and the impacts of a large-scale operational program which may be carried out by the Bureau of Reclamation at the conclusion of the current pilot Colorado Basin Project. Included in the report are the following: a benefit-cost analysis based on current capabilities in snowpack augmentation; an estimate of improvements in the benefit-to-cost ratio to be expected in five years; an assessment of economic, social, environmental impacts in all areas affected; an assessment of the legal and jurisdictional consequences; and a comparison of alternative means for alleviating water problems. The study finds that Congressional action will be needed to allocate water produced by an operational program, to insure that the program is conducted in an equitable manner, and to authorize the program. The report recommends that the program be carried out under a "nonfederal operating authority that would test new institutions and water marketability outside the present public policy environment."[20]

Each of these three studies concludes with positive recommendations regarding the potential of modification in their respective areas. Although these studies necessarily leave unresolved some important questions, as indicated in Chapter II of this report, they constitute initial steps in developing the capacity to assess the broad impacts of weather modification projects. At the same time it should be recognized that the effects of seeding on precipitation amounts are complex and uncertain, and these factors have not been fully accounted for. For this reason, there is danger that these assessments may be considered more definitive than they actually are.

Research Policy

The development of effective weather modification techniques and the success of future operational programs depend upon increased understanding of cloud physics and the interaction of clouds with the larger weather systems in which they are imbedded. These basic research problems can best be investigated by universities and by government laboratories. Imaginative administrative leadership and adequate financial support are needed on a stable basis. The NAS/CAS, in its 1971 report cited above, *The Atmospheric Sciences and Man's Needs: Priorities for the Future*, has defined several areas of critical importance. These are:[31]

a. development of instrumentation for measuring cloud nuclei and cloud particles;

b. laboratory and field experiments to determine specific effects of artificial nucleation on precipitation and other weather modification phenomenon;

c. laboratory and field experiments to extend understanding of the charge-separation mechanism; and

d. determination of the value of remote-sensing techniques for observing changes in atmospheric conditions resulting from weather modification experiments.

Additional areas of research and development suggested in this report include: physical chemistry of nucleating agents, microphysics and dynamics of mesoscale systems, further development of mathematical models, development of remote sensing for severe storm detection, and atmospheric measurement. Clearly, a policy which emphasizes and encourages research is vital to development of an effective national program.

Model Studies

The range of investigations needed to provide the base for technology assessments in the field of weather modification is substantially more complex than is apparent to the layman. The atmosphere has been described by John von Neumann and other respected scientists as the most difficult physical system to understand and to predict. Because of the enormous numbers of degrees of freedom with which the atmosphere can react, the very large extent of the system, and the complex ways in which energy is fed in and out and is transformed within the atmosphere, even very good ideas of excellent scientists can prove to be wrong. Constant testing and up-grading of concepts and theories is essential, and one of the most fruitful procedures has been to develop mathematical models of atmospheric processes and to experiment with them, using electronic computers of high speed and capacity. Models of cumulus clouds can describe the updrafts and the surrounding air flow; models also can describe the growth of droplets and development of precipitation. On a larger scale, models describe the generation of storms and the global circulation. Such mathematical models incorporate specific sets of physical factors or processes, and the interactions of these processes are represented by complex series of operations within the computer. As scientific understanding and computer capability increase, additional processes can be included in the models. As they become more complete, one may hope that the models will simulate the real atmosphere more and more closely. The same kinds of mathematical models permit either the prediction of natural atmospheric behavior, the designing of a weather modification experiment intended to bring about a particular effect, or both. Thus prediction and modification are essentially coupled.

Mathematical models which describe cloud processes, or which account for the interaction of cloud systems and the larger scale weather systems in which they are imbedded, greatly oversimplify the real atmosphere. For this reason model research must be coupled with field research. Because existing research groups are too limited to address the larger scale problems in this integrated way, there is further need for a national laboratory which could marshal the necessary manpower and facilities to develop and maintain a high-quality research effort over a period of years. The two recent NAS reports referred to earlier[6, 31] have emphasized the need for a national laboratory which could manage and carry out theoretical, laboratory, and field programs and could incorporate studies of the social impacts of weather modification.

Techniques of mathematical modeling also may offer a rational, systematic method for dealing with the social impacts of weather modification. Interdisciplinary teams might develop decision-making models, for example, of the hurricane problem for a specific region. An experimental effort along these lines, but in an even more complex field than weather modification, has been made by the Club of Rome Project (Predicament of Mankind) at the Massachusetts Institute of Technology.[33] Here, model studies were conducted to predict the consequences of depletion of natural resources and limited world food production as related to increasing population, industrial output, and pollution. The model has been severely criticized as being simplistic in treating individual factors in the aggregate, in ignoring possible interactions, and in giving little attention to sensitivity of calculations within the model to errors or to approximations. The reliability of the conclusions of the MIT study are highly uncertain, and its pertinence to this study is chiefly that it represents an early step toward a methodology for decision-making which might be applicable in a general sense to weather modification. The test of such a model at this early stage should not be completeness or even accuracy, but whether interactions are handled realistically and whether improvements in the model can be made in a step-by-step manner. As operational programs are developed, it will be necessary also to rely on mathematical models for guidance in program design and for analysis of the results and their impacts.

Regulation

An effective means for regulation of field activities is essential to continued progress in weather modification. Regulation is needed to insure reporting and archiving of accurate data relating to weather modification activities and to avoid contamination of one experimental (or operational) area from another. Without central regulation the effects of weather modification efforts might be vitiated or reversed, and in any case the interpretation of results would be doubtful.

A system of central reporting and regulation also will make it possible to establish a data library which is essential for developing a body of law relating to weather modification activities. Questions of litigation which may relate to the application of tort law, liability, negligence, or other legal concepts or analogies can be answered only through development of a body of law based on the full range of activities for which information can be obtained.

The history of data-reporting of weather modification activities is typical of the field in general. P.L. 85-510 of 1958, in making the NSF responsible for basic research in cloud physics and for research applied to increasing precipitation, also made that agency the focus for collecting and disseminating information on weather modification. NSF proceeded to organize interagency conferences and publish annual reports. In addition, it had the power to require private cloud seeders to report their activities to the government, but it did not enforce this provision until 1966, two years before it was deprived of the authority by new legislation. The lack was partially rectified in 1971, when P.L. 92-205 was enacted requiring that nonfederal weather modification activities be reported to the Department of Commerce.

R. W. Johnson has analyzed the role of federal regulation in managing weather modification activities.[9] His study pointed out that the functions of comprehensive planning, coordination, and project review require flexibility. The Office of Science and Technology (OST) or the Interdepartmental Committee for Atmospheric Sciences (ICAS) were suggested as appropriate responsible agencies for these functions. An administrative structure for systematic regulation was regarded as probably needed by the mid-70s, and the advantages and disadvantages of assigning this function to an existing agency or of creating a new one were discussed. It was suggested that at present the licensing function might be left to the states and separate federal agencies, but that ultimately comprehensive federal licensing may be necessary. The study recommended that a thorough study of indemnification needs in weather modification be carried out. The advantages of diversity of research support and sponsorship were cited; but the view was expressed that as research increases, consideration should be given to assigning major responsibility to a single agency. Finally, operational programs were recognized as appropriate to a wide variety of institutions: federal, state, and local government agencies, and private operators under regulation as necessary. The Johnson study was incorporated in the work of the Task Group on the Legal Implications of Weather Modification, directed by H. J. Taubenfeld.[34] The Task Group recommended that a new Weather Modification Regulatory Board be created to be responsible for supervising, reporting, monitoring, and licensing federal-level operations.

The conclusions of the current study are in large part consistent with those of Johnson's study and the Taubenfeld recommendations. Our conclusion is that a regulatory board or its functional equivalent will be essential if and when substantial operational programs are instituted. In fact, even though present commercial operations constitute only about 5 percent of the total national activity in weather modification, a regulatory board could be established now to administer the Weather Modification Reporting Act. At present the regulatory task would be relatively small, but it would be advantageous to establish regulatory policies and procedures before problems reach an acute state. Establishment of a regulatory board should not be considered to justify, in itself, extension of operational activities.

However, a regulatory board would not provide for all of the important elements of comprehensive management as discussed in this report. Weather modification is an emerging technology which has generated difficult and to some extent new policy issues. Especially is this so in the area of hurricane modification. To determine the potential usefulness of weather modification, emphasis must be placed on scientific research and on assessment of the economic, social, legal, and political impacts.

Administration, Organization, and Funding

During the past two decades efforts in weather modification research have been concerned mainly with scientific and technical problems. A continuing controversy has centered on the distinction between seeding-induced effects and the natural variability of weather. But additional problems related to administration and organization have arisen and today remain a principal constraint on the field. As Oppenheimer and Lambright have pointed out, the technology has advanced faster and more consistently than has the mechanism for its administration and organization.[35]

Prior to 1966 and the publication of NAS report No. 1350, *Weather and Climate Modification: Problems and Prospects,* the general scientific attitude toward weather modification was uncertainty about its efficacy. The guarded optimism expressed in that report marked a turning point and provided the base for growth of research activities since FY1969. Coordination and administration of the federal program has lagged, due partly to the aversion of the scientific community toward administered programs of research, and partly to interagency rivalry. As a consequence, agencies have pursued their individual programs, in some cases with inadequate resources and manpower, and such programs as the drought alleviation activities of 1971 have been planned and carried out without adequate consideration of critical scientific or technical limitations. Efforts of Congress to designate a lead agency have so far failed, as noted in Chapter I of this report.

Efforts of the Interdepartmental Committee for Atmospheric Sciences (ICAS) to strengthen coordination and administration have been partially effective. Following passage of Public Law 90-407, which in 1968 removed NSF's mandate for coordination and data collection, ICAS has conducted an annual Interagency Conference on Weather Modification. It has also recommended increased funding for agency research programs, and in 1971 designated lead agencies for the seven national projects discussed earlier in this report.[10]

Two actions have been taken in the area of data reporting. The National Environmental Policy Act of 1969 (Public Law 91-190) provides that impact statements must be filed with the Council on Environmental Quality for each activity affecting the environment. And Public Law 92-205, passed in 1971, requires that nonfederal weather modification activities be reported to the Department of Commerce.

Federal budgets supporting weather modification research have fallen far below the levels recommended in 1966 by the National Academy of Sciences[5] and the ICAS.[36] However, successive increases of 37 percent and 21 percent occurred in FY1971 and FY1972, as is indicated in Table 3. Further substantial increase which was originally projected for FY1973 has proven to be an overestimate, largely due to

Table 3

Agency Support of Weather Modification—
A Comparative Listing (in Millions of Dollars)

	1965	1966	1967	1968	1969	1970	1971	1972	1973	1974
Actual Agency Support	4.97	7.05	9.92	11.30	10.61	12.02	16.43	19.88[a]	20.37[a]	18.37[a]
Recomm.—NAS 1350 (1966)	4.97	b	b	b	b	30				
Recomm.—Newell Report			11.60			89.56				
Recomm.—NAS (1973)										50

[a] Only small portions of Department of Transportation funds for research on climatic effects of SST are included.
[b] Amounts specified for first and terminal years of period only.

Source: NAS/CAS 1350; NAS/CAS 1973 Report; ICAS 10a; ICAS 16; ICAS 17.

deferral of plans for purchase of aircraft and instrumentation for Project Stormfury. The proposed budget for FY1974 indicates a significant drop in funding for weather modification. Whether this is a temporary anomaly or a reversal of a policy of support for weather modification research is uncertain. In any case it is worth noting that the weather modification program in FY1973 represented about 15 percent of the total meteorological research budget excluding NASA, up from 8 percent of the corresponding figure in 1965.

The largest support for weather modification from 1965 to 1971 has been provided by the Department of Interior, NSF, the Department of Commerce, and the Department of Defense (DOD), in that order. In FY1972 the Department of Transportation initiated a major research program directed at determining the effects on climate of an operational fleet of supersonic transports. Although by far the largest program concerned with inadvertent modification, this is only partially reported by ICAS under modification, and therefore most of it does not appear in Tables 2 and 3. News reports of classified military weather modification activities indicate that the costs reported by the Defense Department are only a fraction of total DOD weather modification expenditures.[12]

Although research funding increased substantially between 1966 and 1972, no agency has been supported at a level which permits a major, sustained effort of the scale required by the fundamental problems. For example, airplanes are indispensable to almost any weather modification program or operational project; yet funds are not available for their procurement. For any of the three major civilian agencies supporting weather modification research (Commerce, Interior, NSF) to acquire an aircraft, explicit Congressional approval must be secured even to release the planes from governmental surplus supplies—the normal source for such equipment.

Table 3 summarizes the funding levels for weather modification since 1965 in comparison with the recommendations of NAS Committee on Atmospheric Sciences in 1966[5] and 1973[6] and of the ICAS Select Panel on Weather Modification in 1966.[36] In conclusion, one may note that the level of actual agency support for weather modification has been lower than that required for the fulfillment of national goals delineated in 1966. The ICAS estimate of necessary funding for FY1970 was almost three times that proposed by NAS 1350 and more than seven times the actual funding level. Although funding has fallen short of recommended levels, the upward trend of the years 1966 to 1972 may be a sign of a more realistic appraisal of the techology's potential. This prospect adds urgency to the need for improved coordination, for assignment of administrative responsibility, and for systematic examination of public policy issues.

V. Existing Mechanisms for Decision-Making

> "Though we may be learned by the help of another's knowledge, we can never be wise but by our own wisdom."
>
> Montaigne

Earlier sections of this report have reviewed some of the critical decisions which have influenced the present program of weather modification. Many more details of the history of the subject are provided in other reports.[9, 37, 38] It must be concluded from these accounts that decisions respecting weather modification have so far been based on unrelated or even contradictory goals. Lacking any objective mechanism for deciding when a specific program of weather modification research should be expanded into an operational stage, this important decision has been left by default to the commercial market or to individual state or federal agencies. No coherent, sustained effort has been exerted to relate the potential contributions of weather modification to social or economic goals, nor even to direct technological development toward identified ends. Like virtually all prior technologies in a less sophisticated era, weather modification has so far been left to develop without explicit consideration of society's needs and values.

The most important decisions of the past ten years or so have been based on initiatives of federal agencies or, more precisely, of individuals or small groups within agencies. Examples are provided by Project Stormfury, initiated by small groups within the Departments of Defense and Commerce, and Project Skywater, initiated by a small group within the Bureau of Reclamation. Other examples could be cited. Agency decisions are strongly influenced by interagency rivalry and the drive for survival, and they stand or fall by Congressional action or inaction on funding. Formal coordination and exchange of information has been provided by the Federal Council for Science and Technology through its subordinate Interdepartmental Committee for Atmospheric Sciences (ICAS). With abolition of the Office of Science and Technology and the Federal Council in early 1973, responsibility for coordination of weather modification will probably be shifted to the National Science Foundation.

States and national governments have made important decisions respecting operational cloud seeding through direct requests to federal agencies for aid in breaking droughts. For example, at the request of the governor of Florida, the National Oceanic and Atmospheric Administration undertook a seeding program to break the south Florida drought of 1971,[39] and similar programs have been inaugurated in other states with participation by government agencies, university groups, and private operators—and, in the Philippine Islands, by the Defense Department.

State legislatures of twenty-nine states have enacted legislation regulating weather modification activities, many of which have proved inconsistent and contradictory as pointed out in Chapter III. Their greatest impact seems to have been to prohibit operational and research field operations in certain states. State authority to grant or withhold operating permits has been the basis for decisions at the local level to block operational programs.[29]

Federal legislation affecting weather modification has been notable chiefly by its passivity. The Clean Air Acts of 1963, 1965, and 1967 are aimed at abatement of air pollution and assert broad national objectives which conceivably could influence weather modification activities, but this appears more a possibility than a reality at present. The Environmental Protection Agency, created in 1970, although an important factor in the field of inadvertent air pollution, has not directly affected deliberate weather modification.

The National Environmental Policy Act of 1969 (P.L. 91-190), establishing the Council on Environmental Quality (CEQ), required the preparation of environmental impact statements (Section 102) for all federal programs significantly affecting the quality of the environment. Under this law impact statements have been filed for weather modification programs, but no provision is made for assessment of complex issues or resolution of controversial viewpoints; these are left to the courts to decide in those cases which are brought to court. The future effect of this Act may be substantial, but it seems unlikely that it can resolve the most serious policy issues discussed in this report. The Weather Modification Reporting Act (P.L. 92-205) passed in December 1971 specifically requires that reports of nonfederal activities be filed with the Department of Commerce. Rules for implementation issued by the Department of Commerce designate the NOAA Office of Environmental Modification as the collection agency. This Act may be the first small step in the federal government's assertion of primary jurisdiction over the states in the area of deliberate weather modification. Although federal agencies, whose activities constitute 95 percent of the total for the nation, are not required to report under this law, an administrative action is under consideration which will extend the requirement to the nonmilitary agencies.

Court actions in a few cases, while resolving specific problems, have demonstrated that the uncertainties associated with weather modification make legal liability very difficult to establish. Since litigation typically arises only when the adverse effects are severe, a shortcoming in the decision-making process is obvious.

Decisions have been influenced by the various studies referred to earlier in this report, which had been initiated and carried out by special groups and individual agencies for limited purposes. There has been no overall planning or coordination and no way for utilizing the studies systematically in the development of a coherent national program.

It is clearly evident that existing mechanisms for decision-making are inconsistent and fragmentary. There is no mechanism for insuring systematic echnological assessments, and responsibility for determination of policy is diffuse and uncertain. The current system for reaching decisions and for administering major research programs is ineffective for even the present weather modification activities. The far more complex problems over the horizon demand a more effective system of management.

VI. Management Alternatives

"Wisdom lies in masterful administration of the
unforeseen."

Robert Bridges

Introduction

The hallmark of successful management in weather modification must
be the capacity to deal with a variety of weather phenomena which
differ markedly in their importance, in their states of development, in
the uncertainties which they involve, and in their social impacts. The
decision-making process in each case should be as simple and direct as
the complexities of the particular phenomenon permit.

In devising a comprehensive system of management for weather
modification it is useful to consider separately the following functions:
(1) technological assessment and examination of issues of public policy,
(2) formulation of policy and its effective implementation, (3)
administrative responsibility for coordination of the total research
program and for direction of certain crucial portions, and (4)
responsibility for management of operational programs if and when
they become justified. Clearly, the options are broader now with
respect to choice of organizational structure and procedures than will
be the case after programs have become more institutionalized. In fact,
rational choice may be possible only at a stage before vested interests
have become entrenched. At the same time, however, it is important to
recognize that conditions will change, so that the determination of
policy and of management structures and procedures must remain
flexible.

67

A few preliminary comments may serve to explain why a variety of policies, structures, and procedures will be needed. In those situations where techniques of weather modification are direct and there is little uncertainty as to prediction of effects, the chief concern should be an accurate assessment of impacts. All affected parties should be identified and heard. Where externalities are minimal, as in clearing of fog from airports, the decision-making process should be uncomplicated and straightforward. In other cases, such as the increasing of orographic precipitation, conflicting interests must be considered, and decision-making may be difficult and lengthy.

In situations where the effects of a proposed program are less certain, as in hurricane modification, decision-making should include specific consideration of the results of possible failure and provision for appropriate liability. Aggressive and sustained research on the social, economic, and legal impacts must undergird any decisions in these more complex programs.

In cases where weather modification efforts are judged to be unwarranted because the uncertainties are too great or the externalities too adverse, interested parties may exert great pressure to influence the decision. Obviously, the policy-determining body must be as objective, strong, and responsive to the public interest as possible. Research should be encouraged on these problems, and as significant new information is determined decisions should be reconsidered.

Technological Assessments and Policy Studies

Thorough, competent, technological assessments and broad examination of policy questions are essential to useful programs in weather modification. The two functions are necessarily so closely linked that they may appropriately be considered together.

Basic policy issues are illustrated by the following examples. Should the federal government attempt to reduce damages due to hurricanes? If precipitation is increased, what compensations should be provided to those affected adversely? At what stage of development should operational programs be encouraged or permitted? What controls should be exercised on large-scale efforts whose results are unknown or highly uncertain? Such issues are to a greater or lesser degree interrelated, and each can be approached only through the results of accurate and thorough technological assessments.

Advice on policy issues in weather modification is needed by the executive branch of the government, by the Congress, and by the general public. No existing body is presently responsible for technological assessments and policy studies, but a number of agencies or institutions offer possibilities for carrying out this function. In the following subsections these are described briefly.

Executive Office of the President. The primary function of the Executive Office (EOP) is to provide information and advice to the President. The resources of the Office are in some respects unique, and it has the capacity to serve as a base for technological assessments and policy studies and for policy initiatives in weather modification. The Council on Environmental Quality (CEQ), part of the Executive Office, is responsible for certain aspects of environmental policy, and until recently the Office of Science and Technology (OST) has had broad responsibilities for policy in science and technology. Currently the EOP has no competence in weather modification, but a group conceivably could be formed which could carry out the needed tasks.

So far the policy problems in weather modification hardly justify development of a special group within the EOP. However, if capabilities in hurricane modification should develop, or if secret military activities should be pursued vigorously, the President will need at least a staff competent to review critically the policy studies carried out elsewhere. Also, the actual carrying out of certain studies within the EOP would offer some advantages: notably, freedom from interagency rivalry, freedom from pressures exerted by the constituencies represented by the agencies, and authority to insure that results of policy studies would receive attention by the agencies. On the other hand, decisions reached within the EOP might tend to be dominated by narrow political considerations, and discussions and decisions might be hidden behind a screen of executive privilege. These weaknesses were illustrated when a report of the President's Science Advisory Committee adverse to SST development was withheld from publication and distribution.

Congressional Office of Technology Assessment. Public Law 92-484, passed in October 1972, authorized a new Office of Technology Assessment (COTA) which will report directly to Congress. This yet-unformed office may eventually play an important role in providing the Congress with independent information and advice on weather modification policy. In comparison with the lead agency or the EOP, the COTA may be expected to be more independent of executive agency interests and power struggles and perhaps more responsive to public interest. Since studies carried out by COTA would presumably be reported publicly, they would serve to inform the public as well as to advise Congress. COTA could not be responsible, of course, for directly advising the executive agencies.

National Advisory Committee on Oceans and Atmosphere (NACOA). In 1971, Public Law 92-125 created this Committee with responsibility for advising the Congress and the President, through the Secretary of Commerce, on progress of the marine and atmospheric sciences and service programs of the United States. The Committee is composed of private citizens with training and experience in many fields relating to marine and atmospheric activities. The NAS Committee on Atmospheric Sciences has recommended that NACOA undertake a major study of the public policy issues of weather modification and of the federal organization and legislation necessary to cope with related problems as they arise.[6] Such a study should help to define these issues more clearly and to point the way toward solutions. The assignment to NACOA of permanent responsibilities for surveillance of weather modification activities would carry with it the following advantages: (1) independence from direct influence or control by Congress or the Executive Branch, (2) responsiveness to a broad segment of the public, and (3) line of communication directly with Congress and the President. The chief disadvantage would be the very limited resources and small full-time staff of the Committee. Whether NACOA can play an influential role in marine and atmospheric policies remains to be seen.

Environmental Institute. In 1970, the Environmental Studies Board of the National Academies of Sciences and Engineering recommended creation of a nongovernmental Institute for Environmental Studies.[40] Concerned with restoring and protecting the quality of the environment, its objectives were only peripherally related to deliberate weather modification. A group convened by the Aspen Institute has discussed a proposal for an Institute for National Alternatives to carry out policy studies on a very wide range of national problems. One component would be concerned with environmental problems including weather modification.

Under sponsorship of the University Corporation for Atmospheric Research (UCAR) during the fall of 1972, representatives of approximately 30 universities discussed creation of an environmental institute. The general consensus was that the purpose of such an institute should be to evaluate scientific, technological, and broad sociological aspects of the problems studied, to formulate alternative policies and alternative strategies for their implementation, and to assess clearly the consequences of the various alternatives. There was agreement that the institute should not *advocate* particular policies or programs, it should not *regulate,* and it should not be involved in administration of research or operations. Essential characteristics were considered to include: high credibility based on the quality of its products, a staff representing a broad range of disciplines and high competence, independence from government control, open dissemination of reports, and continuity of support. It was recognized that the institute must have ready access to decision-makers. Initial support was proposed to be provided by foundations and government. Problems deemed appropriate for study by an environmental institute include basic policies on population, regional energy budgets, strip mining, food contamination, and weather modification.

A nongovernment, nonprofit institute along these lines would offer certain unique strengths. It would be relatively independent of government influence and pressures of vested interests, it could maintain close contacts with highly competent experts in universities and industry, and it would be free to publish and to discuss issues in the public forum. Its chief weakness would be the lack of authority over government (and nongovernment) programs, which would serve to isolate it from governmental decision-making. An environmental institute along the lines considered here probably would be in the strongest position to insure that the broad public interest is adequately represented in technological assessments and policy studies.

National Laboratory. In Chapter IV the role of a national laboratory was described as including technological assessments. The close association of scientists engaged in research with those engaged in assessment should be especially valuable to government agencies in planning research programs; this affords a prime reason for the national laboratory to be located within the lead agency (NOAA).

Executive Department Agencies. Regardless of the location of the laboratory, the lead agency would require a source of competent advice on the broad impacts of weather modification as a basis for program planning. Other agencies also might find it useful to support studies relating to their weather modification missions either in-house or under contract.

Conceivably, technological assessments and policy studies might be carried on by each of the institutions and agencies referred to in the preceding discussion. Their responsibilities are sufficiently distinct that each needs to maintain some cognizance over developing policies in weather modification. The emphasis and degree of comprehensiveness of technological assessments and policy studies should be appropriate to the missions of the various agencies and institutions. For example, in cases in which international issues may be involved, such as hurricane modification or other large-scale activities (e.g., military operations), the special responsibilities and interests of the President, the Congress, the agency managing the program, and the general public will require studies with varying emphases. In a more restricted case, such as hail suppression, the national laboratory might carry out the major studies, subject to review by EOP, COTA, etc.

Policy Decisions

Policy decisions should, of course, be solidly based on studies such as those just discussed. Where the decision should be made, and who should be responsible, must depend upon the issues involved. Two prime candidates for responsibility for policy decisions are the lead agency (NOAA) and the Executive Office of the President. Each probably should be involved, but in somewhat different ways.

In most cases the chief administrative officer of the lead agency (NOAA) is presumably in the best position to understand the complex technical aspects of contemplated programs and to weigh the broad range of possible impacts. However, the formulation of socially responsible public policy is not easy for an agency which may view its own power, prestige, and even survival as depending upon its achievement of clearly visible technical goals. The difficulties of joining together the responsibilities for public policy and for program development are illustrated in the policy decisions of AEC relating to limitations on the development of nuclear weapons and nuclear power, and of NASA in developing objectives for space exploration beyond Apollo. For this reason it will be important that independent evidence relating to technological assessments and policy be provided to the President and the public and also that Congress have its own channel for reliable information on these matters. Thus the executive agencies engaged in weather modification projects should not bear the full responsibility for determining overall policy in that field.

The most difficult policy issues are likely to concern the advisability and the method of attempting to modify hurricanes and other weather phenomena which could seriously affect the lives and welfare of large numbers of people. Decisions relating to such wide-scale potential dangers and disasters must obviously be based on the broadest possible considerations of public policy, but at the same time procedures must be found which permit policy to be implemented effectively. It is the judgment of the authors of this report that the responsibility for decision and implementation should be threefold. (1) The Administrator of NOAA should be in the best position to make immediate decisions; (2) the President will be ultimately accountable; and (3) the Congress should be assured that decisions are based on general policies which it has approved.

At present the Administrator of NOAA reports to the President and to the Congress through the Secretary of Commerce. This structure does not provide the direct lines of authority and responsibility essential to facilitate immediate responses to severe hazards. In other agencies the officers reponsible for weather modification are at lower administrative levels. We conclude, therefore, that responsibility for decisions on mitigation of large-scale weather damages and disasters should be assigned by specific legislation to the Administrator of NOAA, which agency might then either be placed within the proposed Department of Natural Resources or be left within the Department of Commerce.

Guidelines for decisions should be carefully developed through wide public discussion of the issues, through expert and thorough studies, through public hearings, and through interagency coordination. Systematic review might be provided by an advisory panel. By specifying these procedures the legislation would define and limit the nature of the responsibility assigned to the NOAA Administrator and at the same time emphasize the paramount importance of the public interest.

Other policy decisions which may be expected to present special difficulties deal with broad-scale projects where uncertainties are great or where adverse effects are likely to outweigh the benefits. Massive interventions into the processes of precipitation or of energy transfer may present such problems. We cannot be sure that the effects of these efforts would not be very far-reaching or highly deleterious. Programs which apparently have been carried out by the military for hostile purposes in Southeast Asia fall in this category. Only the President can be responsible for preventing such operational programs. Even in this category, however, may be discovered the genesis of useful capabilities, so that research programs should be encouraged, based on thorough, accurate evaluations and on appropriate international assessments and agreements. Obviously, policy must be carefully developed and must be implemented effectively and with certainty.

Some of the policy decisions in weather modification inevitably will be politically sensitive. Those individuals and groups affected adversely, or who perceive themselves in that role, may be expected to oppose even carefully developed decisions. Such issues as whether to increase snowfall in particular regions may have to be worked out finally through the political process. The procedures discussed earlier for broad consideration of impacts, for insuring thorough, objective advice to Congress as well as to the executive agencies, and for open discussion by all affected parties may help to minimize political complications, but they cannot be expected to eliminate them.

Assignment of lead agency responsibility to NOAA and of special responsibilities for decision-making to the NOAA Administrator inevitably will intensify certain interagency political tensions. Much depends upon the caliber of NOAA's leadership, the wisdom of decisions, and the support of the President and the Congress.

Administration of Research Programs

Administration of key field research programs should be the responsibility of the designated lead agency. The importance of this function has been extensively discussed in this report and needs no further elaboration. The various mission agencies should also support and carry on supplementary research directed to their particular needs and responsibilities. Regulation of research activities should be handled by a regulatory board or another body independent of the lead agency or the other mission agencies.

The actual carrying out of research is not considered in detail in this report. As already stated, it should be expected that the national laboratory and other laboratories of NOAA and of other mission agencies would engage in research, and that universities and research institutes would carry out a large portion of the total laboratory and field activities with grant and contract support by the National Science Foundation and other agencies.

Administration of Operational Programs

Operational programs, if and when they are justified, can be administered by the lead agency, by other mission agencies, or by other institutions depending upon the purpose and scope of the program, the availability of manpower and facilities, and other factors. All operations should be subject to federal regulations administered by a regulatory board or other body independent of the agencies responsible for program management.

Alternatively, responsibility for management of operations and of certain limited research programs might be vested in a public corporation created by Congress. Examples are provided by the Communications Satellite Corporation (COMSAT) and the National Railroad Passenger Corporation (AMTRAK). The main advantage of this alternative would be the freedom to employ an industrial system of management instead of civil service. The chief disadvantages would be that a public corporation, relatively remote and insulated from the processes of policy determination, might therefore be unresponsive to public needs and that it would be likely to stress development of operational programs, perhaps prematurely.

Operational programs in the field might be carried on by the mission agencies, by private industry, or by a public corporation under guidelines determined by government. In the variety of operational activities which may develop, any or each of these administrative alternatives may be utilized.

Synthesis

The preceding discussion of the four management functions (technological assessments and policy studies, policy decisions, research administration, and operations administration) is summarized in Tables 4A through 4D. Because weather modification embraces activities ranging widely in scale, urgency, complexity of effects, and level of uncertainty, it has seemed useful to group the various activities into four categories or classes. These are somewhat arbitrary, but they permit useful distinctions to be made, and all important weather modification activities can be subsumed within them. The tables indicate in skeleton form the functions to be carried out by the various participants in management. To some extent the entries represent

Table 4: Participants in Weather Modification Management

A.   *Broad Scale Response to Immediate Disaster*

Example: Reduction of hurricane winds

	Tech. Asses. & Policy Studies	Policy Decision	Admin. of Research	Admin. of Operations
EOP	+	+		
COTA	+			
NACOA	+			
Environ. Inst.	++			
Nat. Lab.	+		+	
Lead Agency*	+	++	++	+
Regulat. Board			0	+
Public Corp.				+

B.   *Broad Scale Response to Chronic Public Need or Opportunity*

Examples: Snowpack augmentation, redistribution of precipitation, drought alleviation

	Tech. Asses. & Policy Studies	Policy Decision	Admin. of Research	Admin. of Operations
EOP	+	+		
COTA	+			
NACOA	+			
Environ. Inst.	++			
Nat. Lab	+		+	
Lead Agency*	+	++	++	+
Mission Agency			+	+
Regulat. Board			+	+
Public. Corp.			0	+

++ Highly Effective or Critically Needed.

 + Effective or needed within special area

 0 Partially Effective

Blank spaces = Inapplicable

 * The entries for Lead Agency assume that the chief executive or administrator of the Lead Agency should be designated as responsible to the President and to the Congress for policy decisions.

Table 4

(Continued)

C. *Broad Scale Irresponsible Proposals*

Examples: Aggressive military activities
Potentially far-reaching adverse effects or great uncertainties

	Tech. Asses. & Policy Studies	Policy Decision	Admin. of Research	Admin. of Operations
EOP	+			
COTA	+			
NACOA	+			
Environ. Inst.	++			
President		++		
Nat. Lab.	+		+	
Lead Agency*	0	+	+	
Regulat. Board			0	

D. *Simple, Direct Applications, Minimal Externalities*

Examples: Cold fog dispersal, hail suppression, lightning suppression

	Tech. Asses. & Policy Studies	Policy Decision	Admin. of Research	Admin. of Operations
EOP	0	+		
COTA	+			
NACOA	+			
Environ. Inst.	+			
Nat. Lab.	+		+	
Lead Agency*	+	+	+	+
Mission Agency	0	0	+	+
Regulat. Board			0	+
Public. Corp.			+	+

alternatives but, as the discussion has emphasized, in most cases responsibilities should be divided among several participants. Some of the agencies and institutions indicated in the tables are now active, some exist but are not now carrying out the designated functions, and some do not exist. The tables apply to a hypothetical future when weather modification is handled in an efficient and responsible manner.

Table 4A concerns future responses to immediate disasters on a broad scale: the possibilities of reducing the maximum winds in hurricanes probably offer the best example. In this case it will be essential to make on very short notice critical decisions involving highly complicated scientific questions and based on complex assessments of benefits and possible losses for large numbers of persons. Important international dimensions also must be weighed.

Table 4A summarizes the conclusions and recommendations of this report concerning activities in the area of immediate disasters. The broad objective technology assessments and policy studies which are essential here can best be carried out by a nongovernmental institute devoted to analysis of environmental problems. Such an institute would provide the unique advantages of independence, objectivity, integrity, and availability of its results to the public. To provide guidance to the Executive Branch and to Congress, certain aspects of these studies also should be carried on by the Executive Office, the Office of Technology Assessment, NACOA, and by the national laboratory and the lead agency. Necessary international negotiations should be handled by the Executive Office of the President and the State Department. Policy decisions could best be made by the Administrator of NOAA, with review by the Executive Office of the President, if appropriate legislation were passed which assigned this responsibility. Administration of research concerned with weather disasters should be the responsibility of NOAA as lead agency in this field. Any operational program developed might be administered by NOAA or by a public corporation under procedures handled by a regulatory board.

Table 4B concerns broad-scale responses to chronic public needs or opportunities. Programs of snowpack augmentation and the redistribution of precipitation appear to be imminent. Already political pressures are being exerted toward drought alleviation. In these cases there is time for careful evaluation prior to decision, although both functions must conform to the time frame of the weather event. For example, in the case of snowpack augmentation the questions of whether to operate, how to operate, and how to minimize or compensate adverse effects would have to be decided before the winter season. Interstate issues and conflicts of special local interests will arise in this category. Assignments of responsibility are similar to those in the disasster area, with the primary distinction that here mission agencies (for example, the Bureau of Reclamation) may be the appropriate agencies for administering research programs or eventual operational programs within procedures set by the regulatory board.

Table 4C concerns proposals for broad-scale projects which, by definition, are irresponsible. Only the President can insure that programs of this type are not carried out, especially by the military. He is therefore specifically indicated here as responsible for policy decisions. COTA should provide information and guidance to the Congress in this area. The lead agency should be expected to review research results and to carry out policy studies in order to identify proposals which might merit shifting to another category.

Table 4D concerns programs for simple, direct applications of weather modification technology under conditions in which minimal externalities can be anticipated. Dispersal of cold fog over airports or highways is already operational, and the supppression of hail and of lightning may develop in the future. Key elements in these activities are a sound economic evaluation, assurance of responsible operations, and monitoring of effects including secondary ones. Operations should be subject to appropriate controls under the regulatory board.

Category D offers a wider variety of feasible alternatives than do the others. Assessments of the technology and of the economic and social impacts can be carried out by NOAA, NACOA, COTA, an environmental institute, and by the national laboratory and mission agencies. Policy decisions might be made by the Executive Office, as well as by NOAA, depending upon their range of application. Administration of research and operations could be handled by a mission agency concerned with a particular type of weather modification, a public corporation created for this purpose, or by NOAA. Again, a regulatory board should insure that specified procedures are followed in planning, operations, and data reporting.

Conclusions

Weather modification is an emerging technology still in its early stages. The full technical potential of the subject is still hidden from us, and it is not possible at this time to foresee in detail what forms of modification will be in the public interest or which may be inimical. It is time, however, that society claim the driver's seat and get firm hold of the steering wheel before weather modification careens farther down the road. We have that opportunity if we but act promptly and wisely.

The most important capability so far demonstrated—namely, to increase or to decrease orographic precipitation by significant amounts on a more or less determinate basis—poses immediate and unavoidable questions of public policy. Pressures to use this capability operationally will grow, and we must be prepared to deal with the policy issues and to manage these programs to insure maximum net social benefits.

Although we do not know accurately how other forms of weather modification will take shape in the future, there is little doubt that developments will continue particularly in cumulus modification, diversion of precipitation, and suppression of severe storm winds, hail, and lightning. Year by year technology and social organization become more critically dependent on weather processes, on floods, freezes, high temperatures, hurricanes, tornadoes, smog, drought, etc., and present understanding indicates that future benefits from weather modification activities can be substantially enlarged and expanded. However, these benefits depend upon the establishment of sound mechanisms for policy determination and management as well as upon further scientific and technical development of the field.

This study has considered the major institutions which participate or might participate in overall management of weather modification activities and the functions which they should carry out in a fully effective program directed to socially useful goals. Alternative management structures have been identified, and qualitative distinctions between them have been pointed out. To achieve an effective, integrated program of research and assessment, certain changes in institutional structure and function are fundamental steps. These changes in no sense provide a blueprint for a fixed management structure for the future; rather they are necessary to assure that options are fully explored at each stage of development and that operational programs in weather modification will be responsive to the real needs of society.

First, a lead agency is needed to coordinate the work of the ten or more agencies involved in research in weather modification and to take responsibility and leadership for mounting a large-scale integrated research program addressed to the critical scientific problems. Designation of NOAA by the Executive Office as lead agency would place the responsibility where the necessary broad competence in atmospheric research already exists.

Second, responsibility must be designated unequivocally for policy decisions which concern activities directly related to the saving of lives or to other critical aspects of the national welfare. Especially in cases requiring immediate responses, such as hurricane modification, key decisions should be made by one responsible directly both to the President and to Congress. This person must be fully aware of the technical aspects, the full range of research results, and the social, economic, and legal impacts associated with weather modification. These requirements could be met through legislation charging the Administrator of NOAA with responsibility for policy decisions in this critical area of activities.

Third, to insure that the results of field experiments and operations are available to the scientific community and to decision-makers, all such data and plans generated by federal agencies as well as by state and local agencies and private industry should be required to be reported to a central collection agency. Such an information bank would also serve to prevent duplication or contamination of one program by another in the planning of field research or operations. Under the provisions of Public Law 92-205 nonfederal activities are now reported to the Office of Environmental Modification of NOAA; the requirement should be extended to include activities of federal agencies as well.

Fourth, federal legislation is needed to establish a basis for resolving interstate and international issues relating to weather modification. Fifth, and critically important, objective and thorough studies of the impacts of weather modification activities and of policy alternatives will be essential on a continuing basis. These studies must be of high quality and must earn the respect of the general public as well as of government decision-makers. They could best be carried out as part of the responsibility of a broad-based nongovernmental institute of environmental studies.

List of References

1. R. C. Gentry, "Hurricane Debbie Modification Experiments," *Science* 168 (1970):473-475.
2. Advisory Committee on Weather Control, *Final Report,* Vol. I and II (Washington, D.C.: U.S. Government Printing Office, 1957).
3. S. M. Greenfield, *et al.,* "A Rationale for Weather Control Research," *Transactions, American Geoph. Union* 43, No. 4 (1962):469-489.
4. National Science Foundation, *Weather Modification, Seventh Annual Report, 1965* (Washington, D.C.: U.S. Government Printing Office, 1966).
5. Committee on Atmospheric Sciences, National Academy of Sciences–National Research Council Publ. 1350, *Weather and Climate Modification: Problems and Prospects* (Washington, D.C., 1966).
6. Committee on Atmospheric Sciences, National Academy of Sciences–National Research Council, *Weather and Climate Modification: Problems and Progress* (Washington, D.C., 1973).
7. Special Commission on Weather Modification, National Science Foundation, *Weather and Climate Modification,* NSF 66-3 (Washington, D.C., 1966).
8. W. R. D. Sewell, ed., *Human Dimensions of Weather Modification,* Department of Geography Research Paper 105 (University of Chicago, 1966).
9. R. W. Johnson, "Federal Organization for Control of Weather Modification," in *Controlling the Weather,* edited by H. J. Taubenfeld (New York: The Dunellen Co., Inc., 1970), pp. 131-180.

10. Interdepartmental Committee for Atmospheric Sciences, Federal Council for Science and Technology—Executive Office of the President, Report No. 15a, *A National Program for Accelerating Progress in Weather Modification* (Washington, D.C., 1971).

11. D. Shapley, "Rainmaking: Rumored Use Over Laos Alarms Arms Experts, Scientists," *Science* 176 (June 16, 1972):1216-1220.

12. S. M. Hersh, "Rainmaking Is Used as Weapon by U.S.," *New York Times* (July 3, 1972).

13. P. W. Mielke, L. O. Grant, and C. F. Chappell, "An Independent Replication of the Climax Wintertime Cloud Seeding Experiment," *J. Appl. Meteor.* 10 (December 1971):1198-1212.

14. Office of Weather Modification, National Oceanic and Atmospheric Administration, Environmental Research Laboratory, *High Plains Precipitation Enhancement Research Project,* Preliminary Planning Document (Boulder, Colo., August 1972).

15. R. H. Simpson and J. S. Malkus, "Experiments in Hurricane Modification," *Scientific American* 211 (1964):27-37.

16. Federal Aviation Administration, Department of Transportation, *Potential Economic Benefits of Fog Dispersal in the Terminal Area,* Pts. I and II (Washington, D.C., 1971).

17. Interdepartmental Committee for Atmospheric Sciences, Federal Council for Science and Technology—Executive Office of the President, Report Nos. 16 & 17, *National Atmospheric Sciences Program, Fiscal Years 1973 & 1974.*

18. D. W. Boyd, R. A. Howard, J. A. Matheson and D. W. North, *Decision Analysis of Hurricane Modification* (Menlo Park, Calif.: Stanford Research Institute, 1971).

19. J. A. Lackner, *et al., Precipitation Modification* (Arlington, Virginia: National Water Commission, 1971).

20. L. Weisbecker, "Technology Assessment of Winter Orographic Snowpack Augmentation in the Upper Colorado River Basin: The Impacts of Snow Enhancement," Vol. I and Vol. II (Menlo Park, Calif.: Stanford Research Institute, 1972, prepared for National Science Foundation), V. I, 38 pp, V. II, 613 pp.

21. W. R. D. Sewell, "Weather Modification: When Should We Do It and How Far Should We Go?" in *Weather Modification:*

Science and Public Policy, edited by R. G. Fleagle (Seattle: University of Washington Press, 1968), pp. 94-104.

22. J. A. Crutchfield, "Economic Evaluation of Weather Modification," in *Weather Modification: Science and Public Policy,* edited by R. G. Fleagle (Seattle: University of Washington Press, 1968), pp. 105-117.

23. W. R. D. Sewell, R. W. Kates, and L. Phillips, "Human Response to Weather and Climate," *Geog. Rev.* 58, No. 2 (1968):262-280.

24. J. H. Sims and D. B. Baumann, "The Tornado Threat: Coping Styles of the North and South," *Science* 176 (June 30, 1972):1386-1392.

25. R. W. Johnson, "The Legal Implications and Justifications," in *Precipitation Modification,* by J. A. Lackner, *et al.* (Arlington, Virginia: National Water Commission, 1971).

26. H. J. Taubenfeld, *Weather Modification—Law, Control, Operations,* National Science Foundation Report No. 66-7 (Washington D. C., 1966).

27. R. J. Davis, *et al., The Legal Implications of Atmospheric Water Resources Development and Management* (College of Law, University of Arizona, 1968).

28. R. S. Hunt, "Weather Modification and the Law," in *Weather Modification: Science and Public Policy,* edited by R. G. Fleagle (Seattle: University of Washington Press, 1968), pp. 118-137.

29. L. J. Carter, "Weather Modification: Colorado Heeds Voters in Valley Dispute," *Science,* 180 (June 29, 1973):1347-1350.

30. Congressional Record: Senate, March 8, 1973 (S 4128-4199). April 12, 1973 (S 7319-7320). Congressional Record: House, March 28, 1973 (H 2225-2226).

31. Committee on Atmospheric Sciences, National Academy of Sciences—National Research Council, *The Atmospheric Sciences and Man's Needs: Priorities for the Future* (Washington, D.C., 1971).

32. R. A. Howard, J. E. Matheson, D. W. North, "The Decision to Seed Hurricanes," *Science* 176 (June 16, 1972):1191-1202.

33. D. H. Meadows, D. L. Meadows, J. Randers, and W. W. Behrens III, *The Limits to Growth,* a report for the Club of Rome's Project on the Predicament of Mankind (New York: Universe Books, 1972).

34. H. J. Taubenfeld, ed., *Controlling the Weather* (New York: The Dunellen Co., Inc., 1970).

35. J. C. Oppenheimer and W. H. Lambright, "Technology Assessment and Weather Modification," *South Calif. Law Rev.* 45, No. 2 (Spring 1972):570-595.

36. Interdepartmental Committee for Atmospheric Sciences, Federal Council for Science and Technology—Executive Office of the President, Report No. 10a, *A Recommended National Program in Weather Modification* (Washington, D.C., 1966).

37. R. G. Fleagle, "Background and Present Status of Weather Modification," in *Weather Modification: Science and Public Policy,* edited by R. G. Fleagle (Seattle: University of Washington Press, 1968), pp. 3-17.

38. G. J. F. MacDonald, "Federal Government Programs in Weather Modification," in *Weather Modification: Science and Public Policy,* edited by R. G. Fleagle (Seattle: University of Washington Press, 1968), pp. 69-86.

39. J. S. Simpson, W. L. Woodley, and R. M. White, "Joint Federal-State Cumulus Seeding Program for Mitigation of 1971 South Florida Drought," *Bull. Am. Meteor. Soc.* 53 (1972):334-344.

40. Environmental Studies Board, National Academy of Sciences—National Academy of Engineering, *Institutions for Effective Management of the Environment* (Washington, D.C., 1970).

Index